The Mathematical Foundation of Informatics

The Mathematical Foundation of Informatics

Proceedings of the Conference

Hanoi, Vietnam 25 – 28 October 1999

Editors

Do Long Van
Institute of Mathematics, Vietnam

M Ito
Kyoto Sangyo University, Japan

World Scientific

NEW JERSEY • LONDON • SINGAPORE • BEIJING • SHANGHAI • HONG KONG • TAIPEI • CHENNAI

Published by

World Scientific Publishing Co. Pte. Ltd.

5 Toh Tuck Link, Singapore 596224

USA office: 27 Warren Street, Suite 401-402, Hackensack, NJ 07601

UK office: 57 Shelton Street, Covent Garden, London WC2H 9HE

British Library Cataloguing-in-Publication Data
A catalogue record for this book is available from the British Library.

THE MATHEMATICAL FOUNDATION OF INFORMATICS
Proceedings of the Conference

ISBN 981-02-4656-0

Printed in Singapore by World Scientific Printers (S) Pte Ltd

Preface

The first international conference organized in Vietnam, which concerns theoretical computer science, was the ICOMIDC Symposium on Mathematics of Computation, held in Ho Chi Minh City in 1988. For the last years great developments have been made in this areas. Therefore, it had become necessary to organize in Vietnam another international conference in this field, which would enable Vietnamese scientists, especially young people, to update the knowledge, to make contacts, to exchange ideas and experiences with leading experts all over the world.

For such a purpose, the conference on Mathematical Foundation of Informatics (MFI99), held at the Institut de Francophonie pour Informatique (IFI) in Hanoi, was co-organized by the Institute of Mathematics and Institute of Information Technology, Vietnam National Center for Natural Sciences and Technologies (now, Vietnam Academy of Science and Technology). This conference was also endorsed as one of the activities of the South East Asian Mathematical Society (SEAMS).

The Program Committee consisted of André Arnold, Jean Berstel, Marc Bui, Robert Cori, Bruno Courcelle, Karel Culik II, Janos Demetrovics, Josep Diaz, Volker Diekert, Phan Dinh Dieu, Dinh Dung, Jozef Gruska, Masami Ito, Helmut Jürgensen, Juhani Karhumäki, Takuya Katayama, Gyula O. H. Katona, Bach Hung Khang, Hoang Kiem, Daniel Krob, Ivan Lavallée, Bertrand Le Sa ec, Igor Litovsky, Maurice Nivat, Dominique Perrin, Dang Huy Ruan, Jacques Sakarovitch, Ludwig Staiger,

Howard Straubing, Ngo Dac Tan (Secretary), Nguyen Quoc Toan, Do Long Van (Chair).

The Steering Committee consisted of Ding Dung, Wanida Hemakul, Bach Hung Khang, Kar Ping Shum, Polly Wee Sy, Dao Trong Thi, Nguyen Dinh Tri, Do Long Van, Tran Duc Van.

The Organizing Committee consisted of Le Tuan Hoa (Chair), Le Hai Khoi, Michel Mouyssinat, Ngo Dac Tan, Le Cong Thanh.

The main sponsors of MFI99 are: UNESCO Jakarta, Vietnam National Program for Basic Research in Natural Sciences, Institut de Francophonie pour Informatique (IFI), Vietnam Union of Science and Technology Associations (VUSTA), and the Institute of Computer Science at Kyoto Sangyo University.

At the conference, invited lectures were delivered by André Arnold, Ho Tu Bao, Jean Berstel, Christian Choffrut, Nguyen Huu Cong, Robert Cori, Bruno Coucelle, Volker Diekert, Nguyen Cat Ho, Dang Van Hung, Masami Ito, Helmut Jürgensen, Juhani Karhumäki, Takuya Katayama, Gyula O. H. Katona, Nguyen Huong Lam, Ivan Lavallée, Bertrand Le Saëc, Igor Litovsky, Maurice Nivat, Jacques Sakarovitch, Kar Ping Shum, K. G. Subramanian, Ngo Dac Tan, Klaus Wagner. Over 40 contributions in different aspects of theoretical computer science were presented at MFI '99.

This volume consists of several invited lectures and selected contributions at MFI '99. The editors thank the members of the Program Committee and also many referees for evaluation of the papers. We are grateful to all the contributors of MFI '99, especially to the invited speakers who have made a very successful and impressive conference.

We would like to express our thanks to the members of the Steering Committee and Organizing Committee for their cooperation and assistance in the preparation process for the conference and during the conference. Sincere thanks are due to the organizations-sponsors without their supports the conference would not be organized.

We would like to thank Prof. Bruno Courcelle for his report on the conference in the bulletin of EATCS as well.

Finally, the editors apologize to the contributors for a long delay in publishing the proceedings volume.

July 2005

Editors:
Do Long Van
Masami Ito

Contents

On Growth Function of Petri Net and its Applications

Pham Tra An
Institute of Mathematics, P.O. Box 631, BoHo, Hanoi, Vietnam

Abstract

In this paper is introduced the growth function of a Petri net. We show that the growth function of any Petri net is bounded by a certain polynomial. There are relations between the growth function and the representative complexity of the language which is accepted by a Petri net. Some applications are examined.

1 Introduction

Petri net was introduced in 1962 by C. Petri, in connection with a theory proposed to model the parallel and distributed processing systems. From then onwards, the theory of Petri net was developed extensively by many authors (see, for example, [10-13]).

In a Petri net, each place describes a local state, and each marking describes a global state of the net. Since the number of tokens which may be assigned to a place can be unbounded, there may be an infinity of markings for a Petri net. From this point of view, a Petri net could be seen as an infinite state machine.

In order to study thus infinite state machines, in this paper we propose a new tool : the notion of state growth speed, which is called to be the growth function of the machine. An analogous growth function for Lindenmayer systems was earlier considered by some authors, (see [2-3]). As we shall see in the sequel, in the theory of growth function, only the state growth speed of the system matters, no attention is paid to the states themselves. This implies that many problems which are very hard for the infinite state machine in general, but could become solvable for the growth function. From the obtained results on growth function of Petri nets, we hope that it could shed a light to some problems concerning with the capacity of Petri nets.

The purpose of this paper is study of growth function of Petri nets and its applications.

The definitions of Petri net and of Petri net language are recalled in Section 2. The Section 3 deals with the notion of growth function of a Petri net. The main result of this part is the growth speed theorem which shows that the growth function of any Petri net is bounded by a certain polynomial. The Section 4 is devoted to the relations between growth function of a Petri net and representative complexity of the language, which is accepted by this Petri net. Finally we close the paper with a remark and an open problem in Section 5.

2 Definitions

We first recall some necessary notions and definitions. For a finite alphabet Σ, Σ^* (resp. Σ^r, $\Sigma^{\leq r}$) denotes the set of all words (resp. of all words of length r, of length at most r)) on the alphabet Σ, Λ denotes the empty word. For any word $\omega \in \Sigma^*, l(\omega)$ denotes the length of ω. Every subset $L \subseteq \Sigma^*$ is called a language over the alphabet Σ. Let N be the set of all non-negative integers and $N^+ = N \backslash \{0\}$.

Definition 1. *A (free-labeled) Petri net* $\mathcal{N}$ is given by a list :

$$\mathcal{N} = (P, T, I, O, \mu_0, M_f),$$

where :

$P = \{p_1, ..., p_n\}$ is a finite set of *places;*
$T = \{t_1, ..., t_m\}$ is a finite set of *transitions* , $P \cap T = \emptyset$;
$I : P \times T \to N$, the *input function;*
$O : T \times P \to N$, the *output function;*
$\mu_0 : P \to N$, the *initial marking;*
$M_f = \{\mu_{f_1}, ..., \mu_{f_k}\}$ is a finite set of *final marking.*

Definition 2. *A marking* μ (global state) of a Petri net $\mathcal{N}$ is a function from the set of places to N :

$$\mu : P \to N.$$

The marking μ can also be defined as a n-vector $\mu = (\mu_1, ..., \mu_n)$ with $\mu_i = \mu(p_i)$ and $|P| = n$.

Definition 3. A transition $t \in T$ is said to be *firable at the marking* μ if :

$$\forall p \in P : \mu(p) \geq I(p, t).$$

Let t be firable at μ and if t fires, then the Petri net $\mathcal{N}$ shall change its state from marking μ to a new marking μ' which is defined as follows :

$$\forall p \in P : \mu'(p) = \mu(p) - I(p,t) + O(t,p).$$

We set $\delta(\mu, t) = \mu'$ and the function δ is said to be the *function of changing state of the net.*

A firing sequence can be defined as a sequence of transitions such that the firing of each its prefix will be led to a marking at which the following transition will be firable. By $\mathcal{F}_{\mathcal{N}}$ we denote the set of all firing sequences of the net $\mathcal{N}$.

We now extend the function δ for a firing sequence by induction as follows

Let $t \in T^*, t_j \in T, \mu$ be a marking, at which tt_j is a firing sequence, then

$$\begin{cases} \delta(\mu, \Lambda) & = & \mu, \\ \delta(\mu, tt_j) & = & \delta(\delta(\mu, t), t_j). \end{cases}$$

Definition 4. *The language acceptable by (free-labeled) Petri net $\mathcal{N}$* is the set

$$L(\mathcal{N}) = \{t \in T^* / (t \in \mathcal{F}_{\mathcal{N}}) \wedge (\delta(\mu_0, t) \in M_f)\}.$$

The set of all (free-labeled) Petri net languages is denoted by $\mathcal{L}^f$.

3 The Growth Function of a Petri Net

3.1. Let $\mathcal{N} = (P, T, I, O, \mu_0, M_f)$ be a Petri net. We denote

$$S_r = \{\mu / (\exists t \in \mathcal{F}_{\mathcal{N}}) \wedge (t \in T^r) \wedge \mu = \delta(\mu_0, t)\},$$

$$S_{\leq r} = \{\mu / (\exists t \in \mathcal{F}_{\mathcal{N}}) \wedge (t \in T^{\leq r}) \wedge \mu = \delta(\mu_0, t)\}.$$

S_r (resp. $S_{\leq r}$) is the set of all reachable markings of $\mathcal{N}$ by firing r (resp. at most r) transitions.

Definition 5. *The Growth Functions $h_{\mathcal{N}}$, $g_{\mathcal{N}}$ of Petri net $\mathcal{N}$* are defined by

$$h_{\mathcal{N}}(r) = |S_r|,$$

$$g_{\mathcal{N}}(r) = |S_{\leq r}|.$$

Now we remark that an exact estimating $g_{\mathcal{N}}(r)$ or $h_{\mathcal{N}}(r)$ will doubtless be a complicated function of r. However, it almost always happens that for large value of r, $g_{\mathcal{N}}(r)$ or $h_{\mathcal{N}}(r)$ can be closely approximated by a much simpler function which will provide us about state growth speed of the net $\mathcal{N}$.

3.2. In the sequel, we use the notations and definitions of the theory of computational complexity.

Definition 6. If f and g are functions defined on the positive integers, then

(1) $f = O(g)$ if there is a $C > 0$ and an $N > 0$ such that $|f(n)| \leq C|g(n)|$ for all $n \geq N$.

(2) $f = \Omega(g)$ if $g = O(f)$.

(3) $f = \Omega(\mathcal{C})$, where $\mathcal{C}$ is a class of functions, if $f = \Omega(g)$ for all $g \in \mathcal{C}$.

The following theorem gives us an upper bound of state growth speed for any Petri net.

Theorem 1. (The growth speed Theorem)

If $\mathcal{N}$ is a Petri net with m transitions and n places, $k = min\{m, n\}$ and P_k is any polynomial of degree k, then

$$h_{\mathcal{N}} = O(P_k),$$

$$g_{\mathcal{N}} = O(P_k).$$

Thus the growth funtion of any Petri net is bounded by a certain polynomial. This is an essential limitation of the Petri net.

Proof. Let $\mathcal{N} = (P, T, I, O, \mu_0, M_f)$ be a Petri net with $|T| = m$, $|P| = n$. We now estimate $|S_{\leq r}|$. There are two ways for doing it.
First we prove $|S_{\leq r}| \leq P_n(r)$ with $|P| = n$.

Denote

$$\mu_0 = (a_1, ..., a_n); \quad a = max\, a_i\ , \quad 1 \leq i \leq n.$$

$$l = max|O(t_j, p_i) - I(p_i, t_j)|, \quad 1 \leq i \leq n; \quad 1 \leq j \leq m.$$

Let $t = t_{j_1} t_{j_2} ... t_{j_p}$, $p \leq r$, be any firing sequence of $\mathcal{N}$. The equation of state change by firing t can be determined as follows :

$\delta(\mu_0, t_{j_1}) = \mu' \quad with \quad \forall p_i \in P:$

$$\mu'(p_i) = \mu_0(p_i) + (O(t_{j_1}, p_i) - I(p_i, t_{j_1})),$$

$$\mu'(p_i) \quad \leq \quad a + l.$$

$\delta(\mu_0, t_{j_1} ... t_{j_p}) = \mu^{(p)} \quad with \quad \forall p_i \in P:$

$$\mu^{(p)}(p_i) = \mu^{(p-1)}(p_i) + (O(t_{j_p}, p_i) - I(p_i, t_{j_p})),$$

$$\mu^{(p)}(p_i) \quad \leq \quad a + p.l \quad \leq \quad a + l.r.$$

Therefore $\forall r \in N^+$:

$$|S_{\leq r}| \leq (a + lr)^n = P_n(r).$$

Second, we show $|S_{\leq r}| \leq P_m(r)$ with $|T| = m$. We define the matrices I^- , O^+ , D as follows :

$$I^-[j,i] = (I(p_i, t_j))_{m \times n}.$$
$$O^+[j,i] = (O(t_j, p_i))_{m \times n}.$$
$$D = O^+ - I^-$$

and set :

$$e[j] = (0, \cdots, 0, \underbrace{1}_{j-th}, 0, \cdots, 0)_{1 \times m}.$$

Let $t = t_{j_1} t_{j_2} ... t_{j_p}$, $p \leq r$, be any firing sequence of $\mathcal{N}$. Firing t, the equation of state change is also determined by another way as follows :

$$\delta(\mu_0, t_{j_1}) = \mu' = \mu_0 + e[j_1]D.$$

$$\delta(\mu_0, t_{j_1} ... t_{j_p}) = \mu^{(p)} = \mu^{(p-1)} + e[j_p]D.$$

We obtain :

$$\delta(\mu_0, t_{j_1} ... t_{j_p}) = \mu_0 + e[j_1]D + ... + e[j_p]D.$$

We set $e[j]D = \nu_j$, $j = 1, \cdots, m$, and f_j is number of occurences of transition t_j in t. We can now express the equation of state change in the following form :

$$\begin{cases} \mu^{(p)} & = & \mu_0 + \sum_{j=1}^m f_j \nu_j, \\ \sum_{j=1}^m f_j & \leq & r. \end{cases}$$

It follows that $|S_{\leq r}|$ equals at most the number of non-negative integer solutions of inequation $\sum_{j=1}^m f_j \leq r$. In [8] we have proved this number equals $C_{m+r}^r = (m+r)!/r!m! \leq (m+r)^m$. Therefore $\forall r \in N^+$:

$$|S_{\leq r}| \leq (m + r)^m = P_m(r).$$

Combining both results of estimating $|S_{\leq r}|$, we obtain :

$$|S_{\leq r}| \leq P_k(r) \,, \quad with \quad k = min\{m, n\}.$$

Finally, from the property $\forall r \in N : |S_r| \leq |S_{\leq r}|$, it follows $|S_r| \leq P_k(r)$, we obtain $h_{\mathcal{N}} = O(P_k)$, $g_{\mathcal{N}} = O(P_k)$. QED.

3.3. We now consider the growth function for some special classes of Petri nets. Denote $\mathcal{S} = \bigcup S_r$, $r \geq 0$. $\mathcal{S}$ is the set of all reachable markings of net.

A Petri net $\mathcal{N}$ is *safe* if $\forall \mu \in \mathcal{S}, \forall p_i \in P : \mu(p_i) \leq 1$, i.e. the number of token in any place is either 0 or 1. Safeness is an important property of hardware devices. If $|P| = n$, then $|\mathcal{S}| \leq 2^n = C$. Therefore for any $r \in N^+$

$$h_{\mathcal{N}}(r) \leq g_{\mathcal{N}}(r) \leq C.$$

A Petri net is *bounded* if there exists a contant K, such that for $\forall \mu \in \mathcal{S}, \forall p_i \in P : \mu(p_i) \leq K$. It is easy to see that if $\mathcal{N}$ is bounded and $|P| = n$, then $|\mathcal{S}| \leq (K+1)^n = C$. Therefore for any $r \in N^+$:

$$h_{\mathcal{N}}(r) \leq g_{\mathcal{N}}(r) \leq C.$$

A Petri net is *conservative* if $\forall \mu \in \mathcal{S}, |P| = n$:

$$\sum_{i=1}^{n} \mu(p_i) = \sum_{i=1}^{n} \mu_0(p_i).$$

Because μ_0 is given, therefore $\sum_{i=1}^{n} \mu_0(p_i) = K$, it implies that $\mu(p_i) \leq K$, i.e. $\mathcal{N}$ is bounded and we obtain also :

$$h_{\mathcal{N}}(r) \leq g_{\mathcal{N}}(r) \leq C.$$

Thus, the growth functions of either safe or bounded or conservative Petri net are bounded by a contant.

4 The Growth Function and Representative Complexity

4.1. In [7-9], we have examined a representative complexity of language, defined as follows :

Let $L \subseteq \Sigma^*$. We define two equivalence relations $E_{\leq r}(modL)$ in $\Sigma^{\leq r}$ (and $E_r(modL)$ in Σ^r) by :

$\forall x_1, x_2 \in \Sigma^{\leq r}$,(and $\forall x_1, x_2 \in \Sigma^r$) :

$$x_1 E_{\leq r} x_2(modL) \Leftrightarrow \forall \omega \in \Sigma^* : x_1\omega \in L \leftrightarrow x_2\omega \in L.$$

$$(x_1 E_r x_2(modL) \Leftrightarrow \forall \omega \in \Sigma^* : x_1\omega \in L \leftrightarrow x_2\omega \in L).$$

It is easy to show that the relations $E_{\leq r}(modL)$, $E_r(modL)$ are reflexive, symmetric and transitive, therefore they are equivalence relations.

We define :

$$G_L(r) = Rank\ E_{\leq r}(modL),$$

$$H_L(r) = Rank\ E_r(modL)$$

where *Rank E* is rank of the equivalent relation E .

They are considered to be representative complexity characteristics of the language L over $\Sigma^{\leq r}$ and over Σ^r. There is a nice relation between the growth functions of a Petri net and the representative complexities of the language which is accepted by this Petri net.

Theorem 2. (The supply-demand Theorem).

Let $L = L(\mathcal{N})$, where $\mathcal{N}$ is a Petri net. Then for any $r \in N^+$

$$H_L(r) \leq h_{\mathcal{N}}(r) + 1,$$

$$G_L(r) \leq g_{\mathcal{N}}(r) + 1.$$

Proof. We first extende the partial function δ to a total function over $T^{\leq r}$ by adding a new marking μ_ϵ defined as follows :

. If x is a firing sequence of $\mathcal{N}$ at μ, then

$$\tilde{\delta}(\mu, x) = \delta(\mu, x)$$

. If x is not a firing sequence of $\mathcal{N}$ at μ, then

$$\tilde{\delta}(\mu, x) = \mu_\epsilon$$

. For all $x \in T^{\leq r}$, $\tilde{\delta}(\mu_\epsilon, x) = \mu_\epsilon$

. Finally $\mu_\epsilon \notin M_f$.

We remark that in a strict sense, μ_ϵ is not a marking, since it is not an n-vector. But here we could consider it to be a special marking of $\mathcal{N}$.

Set $\tilde{S}_{\leq r} = S_{\leq r} \cup \{\mu_\epsilon\}$, and $|\tilde{S}_{\leq r}| = |S_{\leq r}| + 1$.

Now we prove that if $L = L(\mathcal{N})$ then $G_L(r) \leq |\tilde{S}_{\leq r}|$. We assume the contrary that $G_L(r) > |\tilde{S}_{\leq r}|$. There exist $x_1, x_2 \in T^{\leq r}$ such that $x_1 \overline{E}_{\leq r} x_2(modL)$ but $\tilde{\delta}(\mu_0, x_1) = \tilde{\delta}(\mu_0, x_2)$, where $\overline{E}_{\leq r}(modL)$ is the negation of $E_{\leq r}(modL)$. It follows from the last equation that both x_1, x_2 are (or are not) firing sequences and we could verify that :

$$\forall \omega \in T^* : x_1\omega \in L \leftrightarrow x_2\omega \in L.$$

According to the definition, it implies that $x_1 E_{\leq r} x_2(modL)$ which conflicts with hypothesis $x_1 \overline{E}_{\leq r} x_2(modL)$. Therefore :

$$G_L(r) \leq |\tilde{S}_{\leq r}| = |S_{\leq r}| + 1 = g_{\mathcal{N}}(r) + 1.$$

By an analogous argument, we also obtain $H_L(r) \leq h_{\mathcal{N}}(r) + 1$. QED.

4.2. Using the above relation, we get some corollaries and applications.

Corollary 1. *If L is a language with either $H_L = \Omega(P_k)$ or $G_L = \Omega(P_k)$, then L is not acceptable by any Petri net whose numbers of transitions and of places are equal or less than k.*

Proof. In order to prove the corollary, we assume the contrary that L is acceptable by a Petri net $\mathcal{N}$ with $k = min\{|T|, |P|\}$. Applying the theorem 2, and then the theorem 1, we obtain :

$$H_L(r) \le h_{\mathcal{N}}(r) + 1 = O(P_k),$$

$$G_L(r) \le g_{\mathcal{N}}(r) + 1 = O(P_k).$$

This conflics with hypothesis either $H_L = \Omega(P_k)$ or $G_L = \Omega(P_k)$. Therefore L is not acceptable by any Petri net whose numbers of transitions and of places are equal or less than k. QED.

Corollary 2. *If L is a language with either $H_L = \Omega(\mathcal{P})$ or $G_L = \Omega(\mathcal{P})$, where $\mathcal{P}$ is the class of all polynominals, then L is not acceptable by any Petri net.*

Proof. The proof is analogous to the one of corollary 1.

By the Corollaries 1 and 2, we can show a lot of rather simple languages not being acceptable by either any Petri net or a Petri net whose number of transitions and number of places are less than a given contant.

Example 1. Let $|\Sigma| = k \ge 2$, $c \notin \Sigma$ and :

$$L = \{xcx / x \in \Sigma^+\}.$$

It can verify that if $x_1, x_2 \in \Sigma^{\le r}$, $x_1 \ne x_2$ then $x_1 \overline{E}_{\le r} x_2 (modL)$. Therefore $G_L(r) = |\Sigma^{\le r}| = (k^{r+1} - 1)/(k-1) = \Omega(\mathcal{P})$. According to the corollary 2, L is not acceptable by any Petri net.

Example 2. Let $\Sigma = \{0, 1\}$, $c \notin \Sigma$, $k \ge 2$ and :

$$L_k = \{xcx \,/\, x \in \Sigma^*,\ |x|_1 = k\},$$

where $|x|_1$ denotes the number of occurences of 1 in x. We now prove that for any $r \ge k$: $H_{L_k}(r) \ \ge\ P_k(r)$.

We set :

$$W_r \ = \ \{x \,/\, x \in \Sigma^*; l(x) = r; |x|_1 = k\},$$

where $l(x)$ is the length of x. It is easy to show that :

$$|W_r| \ = \ C_r^k = r! \,/\, k!(r-k)! \ = \ r(r-1)\cdots(r-k+1) \,/\, k! \ = \ P_k(r).$$

For any $x_1, x_2 \in W_r$, we prove that if $x_1 \neq x_2$ then $x_1 \overline{E}_r x_2 (modL_k)$. In fact, if we choose $\omega = cx_1$, then $x_1\omega = x_1cx_1 \in L_k$, but $x_2\omega = x_2cx_1 \notin L_k$. It follows $x_1 \overline{E}_r x_2(modL_k)$. Therefore :

$$H_{L_k}(r) \geq |W_r| = P_k(r).$$

According to corollary 1, it implies that L_k is not acceptable by a Petri net whose numbers of transitions and of places are equal or less than k.

Theorem 3. *Let $\mathcal{N}$ be a Petri net with $g_{\mathcal{N}}(r) \leq C$, then $L = L(\mathcal{N})$ is regular.*

Proof. We first recall the Myhill-Nerode's equivalence relation $E(modL)$ defined as follows : $\forall x_1, x_2 \in \Sigma^* \quad :$

$$x_1 E x_2 (modL) \Leftrightarrow \forall \omega \in \Sigma^* \quad : \quad x_1\omega \in L \leftrightarrow x_2\omega \in L.$$

Denote $I_L = Rank\, E(modL)$.

Myhill and Nerode have proved that L is regular if and only if $I_L \leq C$.

From the Theorem 2, $G_L(r) \leq g_{\mathcal{N}}(r) + 1$, it follows that $G_L(r) \leq C$. Because $G_L(r)$ is non-decreasing and bounded, there exists $lim\, G_L(r) = q, \quad q = const$, when $r \to \infty$. Since the values of $G_L(r)$ are integer, so there is a constant r_0, such that $\forall r \geq r_0 : \quad G_L(r) = q$.

For proving L is regular, we assume the contrary that L is not regular. By Myhill-Nerode's theorem, $I_L = +\infty$, therefore there is an infinite sequence $x_1, x_2, ..., x_k, ...$ with $x_i \in \Sigma^*$, $x_i \neq x_j$ and $x_i \overline{E} x_j (modL)$. From this sequence, we pick up the finite sequence $x_1, x_2, ..., x_q, x_{q+1}$ and set $k = Max\{l(x_1), ..., l(x_{q+1})\}$. We now choose $r = Max\{k, r_0\}$. We obtain $x_i \overline{E}_{\leq r} x_j (modL)$ for $i \neq j$. It follows $G_L(r) \geq q+1$. Thus, there is r , $r \geq r_0$ but $G_L(r) \neq q$. This contradicts with the property that $\forall r \geq r_0 , \quad G_L(r) = q$. It follows that L is regular. QED.

Corollary 3. *If Petri net $\mathcal{N}$ has one of following properties : safe, bounded, conservative, then $L = L(\mathcal{N})$ is regular.*

Proof. At the end of Section 1, we have proved that if $\mathcal{N}$ gets one of properties safe, bounded, conservative then its growth functions are bounded. According to theorem 3, it implies that $L(\mathcal{N})$ is regular. QED.

5 Remark and Open Problem

Now we extend the sphere of applying method of growth function.

A (non-erasing) labeled Petri net $\mathcal{N}$ is defined by a list :

$$\mathcal{N} = (P, T, I, O, \sigma, \mu_0, M_f),$$

where P, T, I, O, μ_0, M_f are the sames in Definition 1,

$\sigma : T \to \Sigma$, is a *(non-erasing) labeled function* , where Σ is a finite *output alphabet;*

We can extend the labeled function σ for a sequence as follows :

$$if \quad t = t_1 t_2 ... t_n \quad then \quad \sigma(t) = \sigma(t_1)\sigma(t_2)...\sigma(t_n).$$

The language acceptable by labeled Petri net $\mathcal{N}$ is the set :

$$L(\mathcal{N}) = \{x \in \Sigma^* / \ \exists t \in T^* : (x = \sigma(t)) \wedge (t \in \mathcal{F}_{\mathcal{N}}) \wedge (\delta(\mu_0, t) \in M_f)\}.$$

The set of all labeled Petri net languages is denoted by $\mathcal{L}$.

It is obvious that the free-labeled Petri net is a particular case of labeled Petri net with σ is an isomorphism, then it may be omitted completely by choosing $\Sigma = T$. In [9], we have proved that $\mathcal{L}^f \subset \mathcal{L}$.

Remark. We have proved that the theorems 1 and 2 are still hold for the (non-erasing) labeled Petri net. The result shall be published in the next paper.

Open Problem. Is it possible to apply the method of growth function to other infinite state systems, for example, to the iterative array of finite state automata ? On notions and definitions, concerning iterative array of finite automata, we refer to [4].

Acknowledgement. The author would like to thank the referee for making some valuable suggestions for improving the presentation of the paper.

References

[1] P.D. Dieu, *On a complexity characteristic of languages.* EIK **8** (1972)8/9, 447-460.

[2] A. Salomaa, *On exponential growth in Lindenmayer systems.* Indagationes Mathematical **35** (1973)1, 23-30.

[3] A. Paz and A. Salomaa, *Integral sequential word functions and growth equivalence of Lindenmayer systems.* Information and Control **23** (1973)4, 313-343.

[4] S.N. Cole, *Real-time computation by n-dimentional iterative arrays of finite state machines.* IEEE Trans. Comp. **C-18** (1969)4, 349-365.

[5] M. Jantzen, *Language theory of Petri nets.* LNCS **254** , Springer-Verlag, Berlin, 1987, 397-412.

[6] G. Rozenberg, *Behaviour of elementary net systems.* LNCS **254** , Springer-Verlag, Berlin, 1987, 60-94.

[7] P.T. An, *On a necessary condition for free-labeled Petri net languages.* Proceedings of the Fifth Vietnamese Mathematical Conference, Science and Technics Publishing House, Hanoi, 1999, 73-80.

[8] P.T. An, *A complexity characteristic of Petri net languages.* Acta Mathematica Vietnamica **24**(1999)2,157-167.

[9] P.T. An and P.V. Thao, *On capacity of labeled Petri net languages.* Vietnam Journal of Mathematics **27** (1999)3, 231-240.

[10] W. Brauer, W. Reisig and G. Rozenberg (Eds.), *Petri nets : Central models and their properties.* LNCS **254** , Springer-Verlag, Berlin, 1987.

[11] W. Brauer, W. Reisig and G. Rozenberg (Eds.), *Petri nets : Applications and relationships to other models of concurrency.* LNCS **255** , Springer-Verlag, Berlin, 1987.

[12] G. Rozenberg (Ed.), *Advances in Petri nets 1988.* LNCS **340** , Springer-Verlag, Berlin, 1988.

[13] G. Rozenberg (Ed.), *Advances in Petri nets 1989.* LNCS **424** , Springer-Verlag, Berlin, 1990.

[14] J.L. Peterson, *Petri net theory and the modeling of systems.* Prentice-Hall, New York, 1981.

[15] J.E. Hopcroft and J.D. Ullman, *Introduction to automata theory, languages and computation.* Addison-Wesley, New York, 1979.

On an Infinite Hierarchy of Petri Net Languages

Pham Tra An and Pham Van Thao
Institute of Mathematics, P.O. Box 631, BoHo, Hanoi, Vietnam

Abstract

In this paper we show the existence of an infinite hierarchy of Petri net languages on the number of transitions and places of their recognizing nets.

1 Preliminaries

As well-known, the Petri net is a mathematical model of parallel and distributed computing systems. In the last years, the theory of Petri nets and its applications have been investigated extensively by many authors (see, for example, [8-11]).

Let $\mathcal{N}$ be a Petri net with m transitions and n places, and $k = min\{m, n\}$. For any integer $n \geq 1$ we denote by $\mathcal{L}(n)$ the class of all Petri net languages acceptable by a Petri net with $k \leq n$.

Our aim in this paper is to prove that there exists an increasing infinite sequence of integers n_i,

$$1 \leq n_1 < n_2 < \cdots < n_i < n_{i+1} < \cdots,$$

such that

$$\mathcal{L}(n_1) \subset \mathcal{L}(n_2) \subset \cdots \subset \mathcal{L}(n_i) \subset \mathcal{L}(n_{i+1}) \subset \cdots$$

The proof of the result is based on a complexity characteristic of Petri net languages, obtained earlier by the first author of this note [6].

Analogous hierarchies for some other language classes were earlier considered by several authors, for instance, by Cole for languages recognizable by iterative arrays of finite automata [1], by P. D. Dieu and the first author of this note for languages recognizable by probabilistic automata and those with a time-variant-structure [3-4].

Definitions of Petri nets and Petri net languages are recalled in this section. In Section 2 a complexity charactristic of languages is considered. Using this characteristic a necessary condition for the Petri net languages is given. However as it will be shown, this condition is not sufficient. In Section 3, we show the existence of an infinite hierarchy of Petri net languages on the number of transitions and places of their recognizing nets.

For any finite alphabet Σ, we denote Σ^*, (resp. Σ^r, $\Sigma^{\leq r}$) the set of all words (resp. of all words of length r, of all words of length at most r) on the alphabet Σ, Λ denotes the empty word. For any word $\omega \in \Sigma^*$, $l(\omega)$ denotes the length of ω. Every subset $L \subseteq \Sigma^*$ is called a language over the alphabet Σ. Let N be the set of all non-negative integers and $N^+ = N \backslash \{0\}$.

A *(labeled) Petri net* $\mathcal{N}$ is given by a list :

$$\mathcal{N} = (P, T, I, O, \sigma, \mu_0, M_f),$$

where :
$P = \{p_1, ..., p_n\}$ is a finite set of *places;*
$T = \{t_1, ..., t_m\}$ is a finite set of *transitions* , $P \cap T = \emptyset$;
$I : P \times T \to N$ is the *input function;*
$O : T \times P \to N$ is the *output function;*
$\sigma : T \to \Sigma$ is the *labeling function* , where Σ is a finite *output alphabet;*
$\mu_0 : P \to N$ is the *initial marking;*
$M_f = \{\mu_{f_1}, ..., \mu_{f_k}\}$ is the finite set of *final markings.*
We can extend the labeling function σ for the words in T^* as follows :

$$if \quad t = t_1 t_2 ... t_n \quad then \quad \sigma(t) = \sigma(t_1)\sigma(t_2)...\sigma(t_n).$$

A marking μ (global configuration) of the Petri net $\mathcal{N}$ is a function $\mu : P \to N$ from the set of places P into N. The marking μ can also be represented as an n-vector $\mu = (\mu_1, ..., \mu_n)$ where $\mu_i = \mu(p_i)$ and $n = |P|$. A transition t of $\mathcal{N}$ is said to be *firable at the marking* μ if

$$\forall p \in P : \mu(p) \geq I(p, t).$$

If t is firable at μ then when t fires, the Petri net $\mathcal{N}$ will go into a new marking μ' given by

$$\forall p \in P : \mu'(p) = \mu(p) - I(p, t) + O(t, p).$$

We write then $\delta(\mu, t) = \mu'$ and call δ *the state changing function of the net* $\mathcal{N}$.

A *firing sequence of* $\mathcal{N}$ can be defined as a sequence of transitions such that the firing of each of its prefix will lead N into a marking at which the next transition is firable. The set of all firing sequences of $\mathcal{N}$ is denoted by $\mathcal{F}_{\mathcal{N}}$.

The function δ can be extended for firing sequence by induction as follows

$$\begin{cases} \delta(\mu, \Lambda) & = & \mu, \\ \delta(\mu, tt_j) & = & \delta(\delta(\mu, t), t_j), \end{cases}$$

where $t \in T^*$, $t_j \in T$ and μ is a marking at which tt_j is a firing sequence.

We call *language acceptable by a (labeled) Petri net* $\mathcal{N}$ the set :

$$L(\mathcal{N}) = \{x \in \Sigma^* / \ \exists t \in T^* : (x = \sigma(t)) \wedge (t \in \mathcal{F}_{\mathcal{N}}) \wedge (\delta(\mu_0, t) \in M_f)\}.$$

The set of all (labeled) Petri net languages is denoted by $\mathcal{L}$.

2 On a criterion for Petri net languages

In this section we recall a necessary condition for Petri net languages introduced in [7] (Theorem 2.4) and show that the condition is not sufficient (Theorem 2.8). This condition is based on a complexity characteristic for languages defined as follows :

With every language $L \subseteq \Sigma^*$ we associate an equivalence relation on $\Sigma^{\leq r}$, denoted by $E_{\leq r}(modL)$, and an equivalence relation on Σ^r, denoted by $E_r(modL)$, which are defined respectively as follows

$$\forall x_1, x_2 \in \Sigma^{\leq r}, x_1 E_{\leq r} x_2(modL) \Leftrightarrow \forall \omega \in \Sigma^* : x_1\omega \in L \leftrightarrow x_2\omega \in L;$$

$$\forall x_1, x_2 \in \Sigma^r, (x_1 E_r x_2(modL) \Leftrightarrow \forall \omega \in \Sigma^* : x_1\omega \in L \leftrightarrow x_2\omega \in L).$$

Then we define :

$$G_L(r) = RankE_{\leq r}(modL),$$

$$H_L(r) = RankE_r(modL).$$

With $RankE_{\leq r}(modL)$ is the rank of the equivalent relation $E_{\leq r}(modL)$. They are used as complexity characteristics of the language L. It is easy to see that for any $r \in N$:

$$1 \leq H_L(r) \leq G_L(r) \leq Exp(r).$$

where $Exp(r)$ denotes some exponential function of r.

Let's take some examples :

Example 2.1 Let $\Sigma = \{a, b\}$ and $L_1 = \{a^m b^n / m, n \in N^+\}$.

Consider the subsets

$$W_1 = \{\Lambda\};$$

$$W_2 = \{a^m / 1 \leq m \leq r\};$$

$$W_3 = \{a^m b^k / m + k \leq r; \quad k \geq 1\};$$

$$W_4 = \{\omega \in \Sigma^{\leq r} / \omega \notin W_1 \cup W_2 \cup W_3\}.$$

Obviously $\Sigma^{\leq r} = W_1 \cup W_2 \cup W_3 \cup W_4$ and $W_i \cap W_j = \emptyset$, $i \neq j$

It is easy to prove that any two words in every W_i , $i = 1, 2, 3, 4$, are equivalent however any two words in different sets w_i are not equivalent by the relation $E_{\leq r}(modL_1)$. Therefore $G_{L_1}(r) = 4$.

Example 2.2 Let $|\Sigma| = k \geq 2$ and $L_2 = \{xx^R / x \in \Sigma^*\}$, where x^R is the inverse image of x.

It is easy to show that if $x_1, x_2 \in \Sigma^r$, $x_1 \neq x_2$ then $x_1 \overline{E}_r x_2 (modL_2)$, thereby $H_{L_2}(r) = |\Sigma^r| = k^r$, where $\overline{E}_r$ is the negation of the equivalent relation E_r.

Example 2.3 Let $|\Sigma| = k \geq 2$, $c \notin \Sigma$ and $L_3 = \{xcx / x \in \Sigma^*\}$.

It can be verify that if $x_1, x_2 \in \Sigma^{\leq r}$, $x_1 \neq x_2$ then $x_1 \overline{E}_{\leq r} x_2 (modL_3)$. Therefore $G_{L_3}(r) = |\Sigma^{\leq r}| = (k^{r+1} - 1/(k-1)$.

The following result has been established in [7].

Theorem 2.4 *Let L be accepted by a Petri net with m transitions and n places and $k = min\{m, n\}$. There exists a polynomial P_k of degree k such that, for any integer $r \geq 1$,*

$$H_L(r) \leq P_k(r),$$

$$G_L(r) \leq P_k(r).$$

Using the theorem 2.4, we can show a series of rather simple languages not being acceptable by any Petri net.

Example 2.5 Let $|\Sigma| = k \geq 2$ and $c \notin \Sigma$. Consider the languages $L_2 = \{xx^R / x \in \Sigma^*\}$, $L_3 = \{xcx / x \in \Sigma^*\}$, where x^R is the inverse image of x.

We have proved in examples 2.2 and 2.3 that $H_{L_2}(r) = k^r$ and $G_{L_3}(r) = k(k^r - 1)/(k-1)$. By Theorem 2.4 we have $L_2 \notin \mathcal{L}$ and $L_3 \notin \mathcal{L}$.

Now we shall show that the necessary condition in theorem 2.4 is not sufficient. For this we need some notions in the theory of codes (see [14]).

A language $L \subseteq \Sigma^*$ is *a code over* Σ if $\forall n, m \geq 1$ and $x_1, \cdots x_n, x'_1, \cdots, x'_m \in L$, the equality :

$$x_1 x_2 \cdots x_n = x'_1 x'_2 \cdots x'_m$$

implies $n = m$ and $x_i = x'_i$ for $i = 1, \cdots, n$.

In other words, a set L is a code if any word in L^* can be written uniquely as a product of words in L, that is it has a unique factorization on words of L.

A subset L of Σ^* is a *prefix set* if no word in L is a proper left of another word in L. Evidently every prefix set L with $L \neq \{\Lambda\}$ is a code called a prefix code.

It is not difficult to check that if L is a prefix code then every word $x \in \Sigma^+$ can be written uniquely in the form $x = \tilde{x}x_0$, where $\tilde{x} \in L^*$ and x_0 has no left factor in L.

Using the above fact we obtain

Lemma 2.6 *If L is a prefix code, then for any $r \in N^+$:*

$$G_{L^+}(r) \leq G_L(r),$$

where $L^+ = L^* \backslash \{\Lambda\}$.

Proof. We denote by $\Sigma_1^{\leq r}$ the set of all words of length at most r and no any prefix in L. As $\Sigma_1^{\leq r} \subseteq \Sigma^{\leq r}$, we have $RankE_{\leq r}(modL)$ over $\Sigma_1^{\leq r}$ is not greater than $RankE_{\leq r}(modL)$ over $\Sigma^{\leq r}$ (1).

In the other hand, L is a prefix code, so with $\forall x, y \in \Sigma^{\leq r}$, x and y can be written uniquely in the form $x = \tilde{x}x_0$, $y = \tilde{y}y_0$, where $\tilde{x}$, $\tilde{y} \in L^*$, x_0 , $y_0 \in \Sigma_1^{\leq r}$.

Now we prove that if $x_0 E_{\leq r} y_0(modL)$ then $xE_{\leq r}y(modL^+)$ (2).

Indeed, if $x_0 E_{\leq r} y_0(modL)$ then $\forall \omega \in \Sigma^*$, $x_0\omega \in L \longleftrightarrow y_0\omega \in L$. Two cases are possible :

Case 1 : $x_0\omega \in L$ and $y_0\omega \in L$.

From $x_0\omega \in L$ and $y_0\omega \in L$, it follows $\tilde{x}x_0\omega \in L^+$ and $\tilde{y}y_0\omega \in L^+$, i.e. that $x\omega \in L^+$ and $y\omega \in L^+$, therefore $xE_{\leq r}y(modL^+)$.

Case 2 : $x_0\omega \notin L$ and $y_0\omega \notin L$.

If $x_0\omega \in L^+$, $y_0\omega \in L^+$ then $\tilde{x}x_0\omega \in L^+$ and $\tilde{y}y_0\omega \in L^+$, i.e. $x\omega \in L^+$ and $y\omega \in L^+$.

If $x_0\omega \notin L^+, y_0\omega \notin L^+$ then $\tilde{x}x_0\omega \notin L^+$ and $\tilde{y}y_0\omega \notin L^+$, i.e. $x\omega \notin L^+$ and $y\omega \notin L^+$.

Let $x_0\omega \in L^+$, $y_0\omega \notin L^+$. We have $x_0\omega = (x_0\omega_0)\tilde{\omega} \in L^+$, where $x_0\omega_0 \in L$, $\tilde{\omega} \in L^+$. On other hand $y_0\omega = (y_0\omega_0)\tilde{\omega} \notin L^+$, L is prefix and $\tilde{\omega} \in L^+$, it follows $y_0\omega_0 \notin L$. i.e. exists ω_0 such that $x_0\omega_0 \in L, y_0\omega_0 \notin L$. This conflicts with hypothesis $x_0 E_{\leq r} y_0(modL)$.

Thus in the both cases we have proved that $xE_{\leq r}y(modL^+)$.

From (2), we get $RankE_{\leq r}(modL^+) \leq RankE_{\leq r}(modL)$ over $\Sigma_1^{\leq r}$ (3).

From (1) and (3), we have :

$$RankE_{\leq r}(modL^+) \leq RankE_{\leq r}(modL)$$

This completes the proof.

Now we can establish the main result of this section.

Theorem 2.7 *There exists a language L with $G_L(r) \leq P_1(r)$, which can not be accepted by any Petri net. In other words the necessary condition in Theorem 2.4 is not sufficient.*
Proof. We consider the language :

$$L' = \{a^n b^n | n > 1\}.$$

This language L' is easily verified to be accepted by the Petri net $\mathcal{N}$, described as follows :

$$\mathcal{N} = (\{p_1, p_2, p_3\}, \{t_1, t_2, t_3\}, I, O, \sigma, (1,0,0), \{(0,0,1)\}),$$

where $\sigma(t_1) = a$ and $\sigma(t_2) = \sigma(t_3) = b$, $I(p_1, t_1) = I(p_1, t_2) = I(p_2, t_2) = I(p_2, t_3) = I(p_3, t_3) = O(t_1, p_1) = O(t_1, p_2) = O(t_2, p_3) = O(t_3, p_3) = 1$ and $I(p, t) = O(t, p) = 0$ for any other p and t .

We can show that $G'_L(r) \leq P_1(r)$.

On the other hand, L' is obviously a prefix code.

Put $L = (L')^+$. By the Lemma 2.6 we have

$$G_L(r) = G_{(L')^+}(r) \leq G_{L'}(r) \leq P_1(r).$$

As shown in [13] by Peterson the language $L = (L')^+$ is not a Petri net language. The Theorem is proved.

3 An infinite hierarchy of Petri net languages

Basing on the Theorem 2.4 we can obtain the solution of problem on infinite hierarchy of Petri net languages :
Theorem 3.1 *There exists an increasing infinite sequence of integers n_i,*

$$1 \leq n_1 < n_2 < \cdots < n_i < n_{i+1} < \cdots$$

with $n_{i+1} = 3n_i + 6$, such that :

$$\mathcal{L}(n_1) \subset \mathcal{L}(n_2) \subset \cdots \subset \mathcal{L}(n_i) \subset \mathcal{L}(n_{i+1}) \subset \cdots$$

Proof. Let $\Sigma = \{0, 1\}$, $c \notin \Sigma$, $k \geq 2$. Consider the language:

$$L_k = \{xcx \, / \, x \in \Sigma^* \, , \, |x|_1 = k\},$$

where $|x|_1$ denotes the number of occurrences of 1 in x.

We now prove two following propositions :

(i) For any $r \geq k$: $H_{L_k}(r) \geq P_k(r)$, therefore $L_k \notin \mathcal{L}(k-1)$.

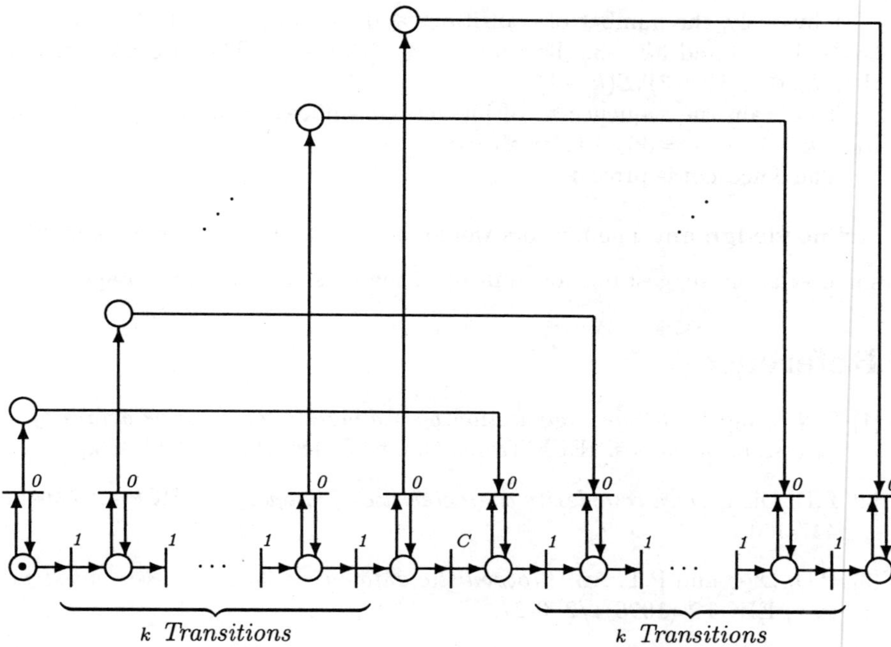

Fig. 1:

(ii) $L_k = L(\mathcal{N})$, where $\mathcal{N}$ is a Petri net with $min\{|T|, |P|\} = 3k + 3$, therefore $L_k \in \mathcal{L}(3k + 3)$.

Put :

$$W_r = \{x \,/\, x \in \Sigma^*; l(x) = r; |x|_1 = k\},$$

where $l(x)$ denotes the length of x. It is easy to verify that :

$$|W_r| = C_r^k = r! \,/\, k!(r-k)! = r(r-1)\cdots(r-k+1) \,/\, k! = P_k(r).$$

For any $x_1, x_2 \in W_r$ with $x_1 \neq x_2$, by choosing $\omega = cx_1$ we have $x_1\omega = x_1cx_1 \in L_k$ whereas $x_2\omega = x_2cx_1 \notin L_k$, that is $x_1\overline{E}_r x_2\,(modL_k)$. This means that

$$H_{L_k}(r) \geq |W_r| = P_k(r).$$

By Theorem 2.4 it follows that $L_k \notin \mathcal{L}(k-1)$.

On the other hand, the language L_k is easily verified to be accepted by the Petri net $\mathcal{N}$, depicted in the above Fig 1. with $\mu_0 = (1, 0, \cdots, 0, 0)$ and $M_f = \{\mu_f = (0, 0, \cdots, 0, 1)\}$.

Obviously, the number of transitions and that of places of $\mathcal{N}$ are respectively $4k+3$ and $3k+3$. Thereby $L_k \in \mathcal{L}(3k+3)$. Thus we have proved that $L_k \in \mathcal{L}(3k+3)\backslash\mathcal{L}(k-1)$.

To obtain the sequence n_i of integers, it suffices to fix a $k \geq 2$ and put $n_1 = k-1$, $n_{i+1} = 3n_i + 6$ for all $i \geq 1$.

The Theorem is proved.

Acknowledgment The authors would like to thank the referee for making some valuable suggestions for improving the presentation of the paper.

References

[1] S.N. Cole, *Real-time computation by n-dimentional iterative arrays of finite state machines.* IEEE Trans. Comp. **C-18** (1969)4, 349-365.

[2] P.D. Dieu, *On a complexity characteristic of languages.* EIK **8** (1972)8/9, 447-460.

[3] P.D. Dieu and P.T. An, *Probabilistic automata with a time-variant structure,* EIK **12** (1976)1/2, 3-27.

[4] P.T. An, *Some necessary conditions for the class of languages accepted by probabilistic automata with a time-variant structure,* EIK **17** (1981)11/12, 623-632 (in Russian).

[5] P.T. An, *On a necessary condition for free-labeled Petri net languages.* Proceedings of the Fifth Vietnamese Mathematical Conference (1999), 73-80.

[6] P.T. An, *A complexity characteristic of Petri net languages.* Acta Mathematica Vietnamica **24** (1999)2, 157-167.

[7] P.T. An and P.V. Thao, *On capacity of labeled Petri net languages.* Vietnam Journal of Mathematics **27** (1999)3, 231-240.

[8] W. Brauer, W. Reisig and G. Rozenberg (Eds.), *Petri nets :Central models and their properties.* LNCS **254**, Springer-Verlag, Berlin, 1987.

[9] W. Brauer, W. Reisig and G. Rozenberg (Eds.), *Petri nets : Applications and relationships to other models of concurrency.* LNCS **255**, Springer-Verlag, Berlin, 1987.

[10] G. Rozenberg (Ed.), *Advances in Petri nets 1988.* LNCS **340**, Springer-Verlag, Berlin, 1988.

[11] G. Rozenberg (Ed.), *Advances in Petri nets 1989.* LNCS **424**, Springer-Verlag, Berlin, 1990.

[12] J.E. Hopcroft and J.D. Ullman, *Introduction to automata theory, languages and computation.* Addison-Wesley, New York, 1979.

[13] J.L. Peterson. *Petri net theory and the modeling of systems.* Prentice-Hall, New York, 1981.

[14] J. Berstel and D. Perrin, *Theory of codes.* Academic Press, New York, 1985.

ALGORITHMS TO TEST RATIONAL ω-CODES

XAVIER AUGROS AND IGOR LITOVSKY
Laboratoire I3S bât. ESSI,
930, Route des Colles, B.P. 145, 06903 Sophia-Antipolis Cedex, France
E-mail: {augros,lito}@i3s.unice.fr

In this paper we present some algoritms to decide whether a given rational language is an ω-code. Those algorithms have a complexity in the worst case in $O(n^3)$, n being the number of states of the automaton representing the language.

1 Introduction

Let Σ be a finite alphabet. We denote by Σ^* (respectively Σ^ω) the set of words (resp. infinite words) over Σ. Given L of Σ^*, the language L^* is the submonoid generated by L and it denotes the set of words factorizable over L. The ω-power L^ω denotes the set of infinite words factorizable over L. A language $C \subseteq \Sigma^*$ is a code (respectively an ω-code[1]) if every word of C^* (respectively C^ω) has a unique factorization over C. Any ω-code is a fortiori a code. Algorithms to test if a given language L is a code can be found in [2], [3] or [4] if the language L is a finite set, and in [5], [6] or [7] for L rational.
To test ω-codes, we present three tests and we give their complexities in time ($O(n^3)$). The first test of the section 4 is based on the interlaced product of automata, the second one[8] and the third one are based on the computation of the left quotients of automata.

2 Preliminaries

Let Σ be a finite alphabet. The set Σ^* is the set of all finite words over Σ, Σ^ω is the set of all infinite words, and Σ^∞ is the union of Σ^* and Σ^ω . The empty word is denoted by ε and $\Sigma^+ = \Sigma^* \setminus \{\varepsilon\}$. Words of Σ^+ are obtained by finite concatenation of letters of Σ: $u = u_1u_2\dots u_n \in \Sigma^+$, for $n > i > 0$, $u_i \in \Sigma$. Words of Σ^ω are obtained by infinite concatenation of letters of Σ: $u = u_1u_2\dots u_n\dots \in \Sigma^\omega$, for $n > i > 0$, $u_i \in \Sigma$.
Let $L \subset \Sigma^*$ be a language over Σ, then L^* is the set of finite words obtained by finite concatenation of words of L, that is the submonoid generated by L, and L^ω is the set of infinite words (also called ω-words) obtained by infinite concatenation of words of L.
We denote by L^∞ the set of all finite or infinite words generated by L: $L^\infty = L^* \cup L^\omega$.
An automaton $\mathcal{A}(L)$ which recognizes a rational language L is a 5-tuple

$(Q, I, F, \delta, \Sigma)$ where Q is the set of states, $I \subseteq Q$ and $F \subseteq Q$ are respectively sets of initial and final states, and δ is the transition function mapping $Q \times \Sigma$ to Q (see [9] for example). $\mathcal{A}$ recognizes a word w if w is the label of a path from an initial state to a final state in $\mathcal{A}$. For infinite words, we give two acceptances modes by an the automaton. The first one, with the Büchi Criterion(see [10]) says that an ω-word w is recognised by an automaton $\mathcal{A} = (Q, I, F, \delta, \Sigma)$, if and only if w is the label of an infinite path reaching infinitely some "final" states of F. In the second one (the Muller mode[10]), a word w is accepted by an automaton $\mathcal{A} = (Q, I, \mathcal{T}, \delta, \Sigma)$, where $\mathcal{T}$ is a subset of the set of the part of Q, if and only if w is the label of path reaching infinitely exactly the states of a set in $\mathcal{T}$.

$\mathcal{A}$ is deterministic if and only if for each state there is a unique transition on each symbol. $\mathcal{A}$ is unambiguous if and only if for every word w recognized by $\mathcal{A}$ there is a unique successful path in $\mathcal{A}$ labeled by w.
$\mathcal{A}$ is normalized if there is exactly one initial state and one final state, and the initial state has no ingoing transition and final state has no outgoing ones. A normalized automaton can be chosen unambiguous.
For any submonoid M, the *root* of M is defined by $Root(M) = (M \setminus \{\varepsilon\}) \setminus (M \setminus \{\varepsilon\})^2$.
Let u and v be two words of Σ^*, $u \sim_L v$ if and only if for all $w \in \Sigma^*$ and $w' \in \Sigma^*$, $wuw' \in L \iff wvw' \in L$.
Let u and v be two words over Σ. We denote by $u < v$ the fact that the word u is a prefix of the word v. It follows that $v = uu'$ for some word u' in Σ^*, the word u' stands for $u^{-1}v$ and is a suffix of the word v.

We denote by pref(v) (respectively suff(v)) the set of all words that are prefixes (resp. suffixes) of the word v, pref(L) (resp. suff(L)) is the set of prefixes (resp. suffixes) of words of L. For two languages L and L', the left quotient of L' by L is $L^{-1}L' = \{w \in \Sigma^* / \text{ for some } u \in L,\ uw \in L'\}$.

A *factorization* of a word u in L^+ (respectively in L^ω) is a finite sequence (resp. an infinite sequence) $f_u = (u_1, u_2, \ldots, u_n, \ldots)$ of words of L such that $u = u_1 u_2 \ldots u_n \ldots$.

We call *sequence of left factors of u for the factorization* f_u, the sequence $(p_i)_{i=1}^n$, where n is the number of factors in the factorization f_u (infinite for infinite words), where for $i \geq 1$, p_i is the concatenation of the i first factors of u: $p_i = u_1 u_2 \ldots u_i$.
Let u, v be two comparable words with respect to prefix order, $f_u = (u_1, \ldots u_n)$ and $f_v = (v_1, \ldots v_m)$ be two factorizations of u and v such that $u_1 \neq v_1$. Let $(p_i)_{i=1}^n$ be the sequence of left factors of u for the factorization f_u and $(q_i)_{i=1}^m$ be the sequence of left factors of v for the factorization f_v. A word s is called a *shift* if there exist two integers, $k \geq 1$ and

$l \geq 1$, such that $s = p_k^{-1}q_l$ (for $k = n$, $p_k \leq q_l$ and for $k < n$, $p_k \leq q_l < p_{k+1}$), or $s = q_l^{-1}p_k$ (for $l = m$, $q_l \leq p_k$ and for $l < m$, $q_l \leq p_k < q_{l+1}$). We denote by $s(f_u, f_v)$ the sequence of shifts for the two factorizations of comparable words u and v, for example on Fig. 1, $s(f_u, f_v) = (s_1, s_2, s_3, s_4, s_5, s_6, s_7)$. $s_i(f_u, f_v)$ denotes the ith shift of (f_u, f_v) (for example on Fig. 1, the 4th shift

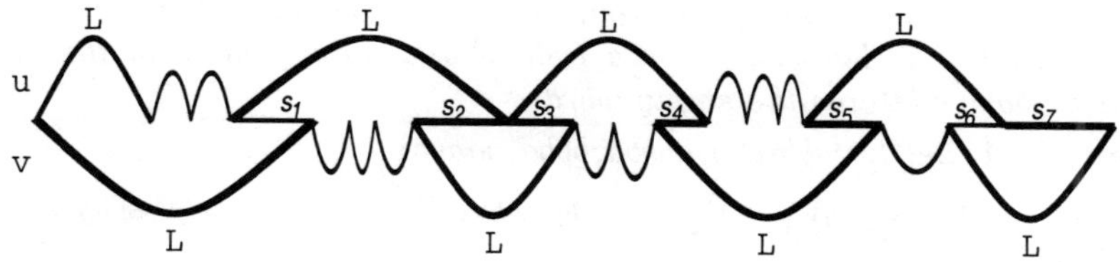

Figure 1. $s(f_u, f_v) = (s_1, s_2, s_3, s_4, s_5, s_6, s_7)$

for the factorizations f_u and f_v is $s_4(f_u, f_v) = s_4$).
A word u is ambiguously covered by L if u has two factorizations with different first factors over L [11].

A language C is a code (see [12] for example) if and only if

$$C^{-1}C \cap C^*(C^*)^{-1} = \{\varepsilon\} \tag{1}$$

in other words C is a code if and only if each finite word of C^* has only one factorization over C, and C is an ω-code if and only if

$$C^{-1}C \cap C^\omega(C^\omega)^{-1} = \{\varepsilon\} \tag{2}$$

i.e. C is an ω-code if and only if each infinite word of C^ω has only one factorization over C [1]. Clearly, any ω-code is a code. A generalization of ω-code, called strict code, has been proposed by Do Long Van in [13]. A strict code is a subset of Σ^∞ such that each word and each infinite word has only one factorization. If $C \subseteq \Sigma^+$, then C is a strict-code if and only if C is an ω-code.
A language C is a code with bounded deciphering delay [12] if it satisfies the following property:

$$\exists d \geq 0\ \forall u \in C\ \forall v \in C\ (uC^d\Sigma^* \cap vC^* \neq \emptyset \Rightarrow u = v) \tag{3}$$

3 Testing unique decipherability of finite words

A.A. Sardinas and C.W. Patterson have proposed a test for codes that allows us to decide if a rational language is a code or not.

3.1 The Test of Sardinas and Patterson

Let us consider the next sequence of sets constructed with a language $L \subset \Sigma^*$:

$$U_1 = L^{-1}L \setminus \{\varepsilon\}$$
$$U_{n+1} = L^{-1}U_n \cup U_n^{-1}L$$

Then we have the following theorem that gives a criterion for codes.
Theorem 1 [5] *A language L is a code if and only if none of the sets U_n defined above contains the empty word.*
Example 1 *Let $L = \{b, abb, abbba, bbba, baabb\}$,*

$$U_1 = \{ba, bba, aabb\}\ , \quad U_2 = \{a, ba, abb\}\ , \quad U_3 = \{a, \varepsilon, bb, bbba, abb, ba\}$$

$\varepsilon \in U_3$, *L is not a code.*

Moreover, this property is decidable for rational languages.
Proposition 1 *If L is a rational language, then each U_n is a rational language and the number of the sets U_n $(n \geq 1)$ is finite.*
Remark 1 *For the test of Sardinas/Patterson, $U_i = \varnothing$ for some $i \geq 1$ if and only if L is a code with bounded deciphering delay* [12].

Example 2 *Let $L = \{ab, abb, baab\}$,*

$$U_1 = \{b\}\ , \quad U_2 = \{aab\}\ , \quad U_3 = \emptyset$$

$U_3 = \emptyset$, *L is a code with bounded deciphering delay.*

Remark 2 *For rational finitary languages, codes with bounded deciphering delay are included in the set of rational ω-codes.*
Remark 3 *For finite finitary languages, codes with bounded deciphering delay are exactly finite ω-codes.*

3.2 A Test by product of automata

Let L be a rational language, let $\mathcal{A}$ be a unambiguous normalized automaton which recognize L. The interlaced product [7] of $\mathcal{A} = (Q, q_0, q_f, \delta, \Sigma)$ by itself is defined as:

$$\mathcal{A}' = \mathcal{A} \times \mathcal{A} = (Q \times Q, (q_0, q_0), (q_f, q_f), \delta', \Sigma).$$

The initial state is (q_0, q_0), the final state is (q_f, q_f).
The transitions δ' are defined as :

$$\delta'((p, q), a) = (p', q')$$

if and only if

$$\delta(p,a) = p',\ \delta(q,a) = q' \text{ if } ((p',q') = (q_f,q_f) \text{ or } (p' \neq q_f \text{ and } q' \neq q_f)),$$
or $\delta(p,a) = p'$, $p' \neq q_f$ and $\delta(q,a) = q_f$, $q' = q_0$,
or $\delta(p,a) = q_f$, $p' = q_0$ and $\delta(q,a) = q'$, $q' \neq q_f$,
remove transitions $\delta'((q_0,q_0),a) = (q_f,q_f)\ \forall a \in \Sigma$.

This automaton recognize some words of L^* which have at least two factorizations over L. The following proposition gives us a criterion to decide if a given language L is a code.

Proposition 2 [7] *A rational language L is a code if and only if the set recognized by the previously defined automaton is empty.*

Example 3 *Let $L = \{a, ba, ca, abac\}$, an normalized antomaton $\mathcal{A}$ is :*

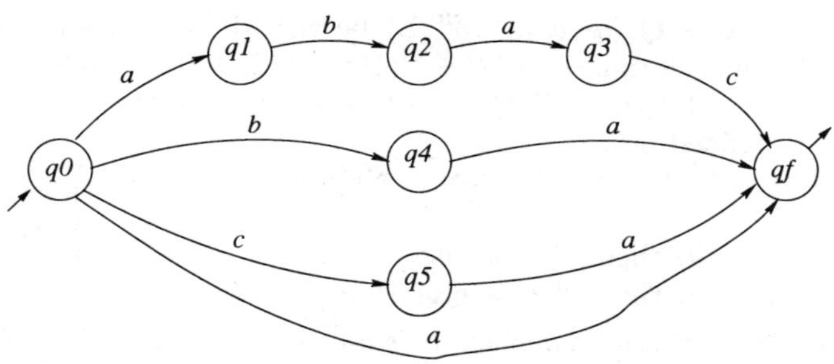

A part of the interlaced product $\mathcal{A} \times \mathcal{A}$ is :

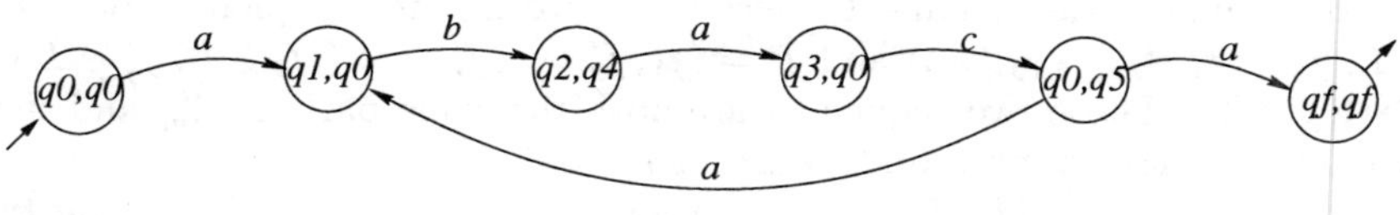

There is a path from (q_0,q_0) to (q_f,q_f) reading the word abaca, for example, which has two factorizations : (a,ba,ca) and $(abac,a)$. The language L is not a code.

Remark 4 *The label of the path between two consecutives states (q_i,q_0) and (q_j,q_0) $(i \geq 0, j \geq 0)$ (or (q_0,q_i) and (q_0,q_j)) is a word of L. for example: the path between (q_1,q_0) and (q_3,q_0) is labeled by $ba \in L$. In other words, we can read on the graph of the automaton $\mathcal{A}'$ the factorizations of the words* *ambiguously covered over L.*

4 Testing rational ω-codes

4.1 A Test by product of automata

With almost the same notations as in Section 3.2, one can decide if a given language L is an ω-code. In Section 3.2 the criterion was the existence of a path, in the automaton defined by the interlaced product $\mathcal{A}' = \mathcal{A} \times \mathcal{A} = (Q \times Q, (q_0, q_0), (q_f, q_f), \delta', \Sigma)$, from the initial state to the final one. To test if a rational language L is an ω-code, the problem is somewhat different. L is not an ω-code if and only if there exists an infinite word, in L^ω, ambiguously covered by L. In an automaton $\mathcal{A}'_m$, constructing from $\mathcal{A}$, such a word is the label of an infinite path begining at the state $(q0, q0)$ and crossing infinitely many particular states.

Let $\mathcal{A}'_m = (Q \times Q, (q_0, q_0), \mathcal{T}, \delta'', \Sigma)$ be the Muller automaton defined from $\mathcal{A}' = \mathcal{A} \times \mathcal{A}$ such that $\mathcal{T} = \{P \in \mathcal{P}(Q \times Q) / \exists i \geq 0 \textit{ and } j \geq 0,\ (q_i, q_0) \in P \textit{ and } (q_0, q_j) \in P\}$ (where $\mathcal{P}(Q \times Q)$ is the set of of the part of the set $Q \times Q$). δ'' is defined by :

$$\left.\begin{array}{l} \delta''((p,q),a) = (p'',q'') \text{ from } \delta'((p,q),a) = (p',q') \\ \text{and} \\ \quad \text{if } p' = q_f \text{ then } p'' = q_0 \text{ else } p'' = p' \\ \quad \text{if } q' = q_f \text{ then } q'' = q_0 \text{ else } q'' = q' \end{array}\right\} (*)$$

Proposition 3 *A rational language L is an ω-code if and only if the Muller autonaton $\mathcal{A}'_m$ do not recognize any word.*

proof. If there exists an ω-word $\dot{w}$ recognized by $\mathcal{A}'_m$, it is the label of an infinite path in $\mathcal{A}'_m$ that infinitly reach some states (q_i, q_0) and (q_0, q_j) (for some $q_i \in Q, q_j \in Q, i \geq 0, j \geq 0$). If there exists a such infinite path in $\mathcal{A}'_m$ then, like in the remark 4, there exist two infinites sequences of words of L, $f = (f_1, f_2, \ldots, f_n \ldots)$ and $f' = (f'_1, f'_2, \ldots, f'_m \cdots)$ ($f_i \in L$, $f'_j \in L$ for $i \geq 1, j \geq 1$). Those two sequences are both the same path in $\mathcal{A}'_m$ and then two factorizations of an infinite word of L^ω.

Conversely, let $f = (f_1, f_2, \ldots, f_n \ldots)$ and $f' = (f'_1, f'_2, \ldots, f'_m \ldots)$ be two factorizations of an ambiguously covered word w. Each factors f_i and f'_j are words of L which are the labels of paths in the automaton $\mathcal{A}$. So to each factors f_i, we can associate a finite sequence (the path of the word f_i in the automaton $\mathcal{A}$): $q_0 \overset{u_{1i}}{\to} q_{1i} \overset{u_{2i}}{\to} q_{2i} \ldots q_{ki} \overset{u_{(k+1)i}}{\to} q_f$ and $f_i = u_{1i} u_{i2} \ldots u_{(k+1)i} \in L$, $u_j \in \Sigma$, $q_{ji} \in Q$ for i and $j \geq 0$.

So for each factorizations f and f', we have the following sequences of paths in the automaton $\mathcal{A}$ (each path being labeled by a factor respectively of f or f'):

$f = (q_0 \overset{u_1}{\rightarrow} q_1 \overset{u_2}{\rightarrow} q_2 \dots q_k \overset{u_{k+1}}{\rightarrow} q_f, q_0 \overset{u_{k+2}}{\rightarrow} q_{k+2} \dots \overset{u_{k+m}}{\rightarrow} q_{k+m} \dots)$ and
$f' = (q_0 \overset{u_1}{\rightarrow} q'_1 \overset{u_2}{\rightarrow} q'_2 \dots q'_k \overset{u_{k+1}}{\rightarrow} q'_{k+1} \overset{u_{k+2}}{\rightarrow} q'_{k+2} \dots \overset{u_{k+m}}{\rightarrow} q_f, q_0 \overset{u_{k+m+1}}{\rightarrow} \dots))$

Now, let us assume that we can apply the rules (*) to those paths. The following path is a part of the result of this operation:
$(q_0, q_0) \overset{u_1}{\rightarrow} (q_1, q'_1) \overset{u_2}{\rightarrow} (q_2, q'_2) \dots (q_k, q'_k) \overset{u_{k+1}}{\rightarrow} (q_0, q'_{k+1}) \overset{u_{k+2}}{\rightarrow} (q_{k+2}, q'_{k+2}) \dots \overset{u_{k+m}}{\rightarrow} (q_{k+m}, q_0) \dots$
This is a path in the automaton $\mathcal{A}'_m$ and it reach infinitely a set of states which contains some states (q_0, q) and (q', q_0) (for q and q' in Q), then the label of this path is an ω-word recognized by $\mathcal{A}'_m$. □

Example 4 *Let $L = \{a, b\} \cup ab^*c$, the following automaton recognizes this language L :*

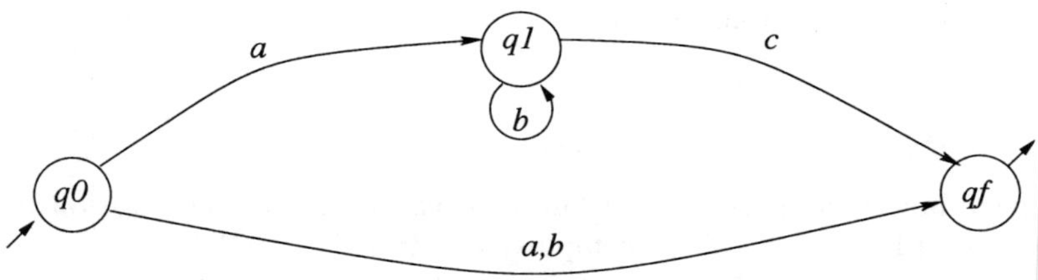

The automaton obtained by the interlaced product $\mathcal{A}' = \mathcal{A} \times \mathcal{A}$ is :

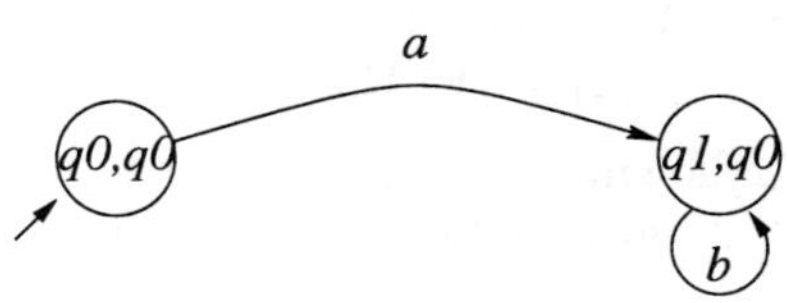

The language L is an ω-code because the single infinite path in $\mathcal{A}'$ cross infinitely only the state (q_1, q_0).

4.2 A Test for strict-codes

It has been proposed by Nguyen Huong Lâm and Do Long Van in [8] a procedure to test if a given language of Σ^∞ is a strict-code.

Let us consider a language L of Σ^∞ and the sequence of sets:

$$U'_1 = L^{-1}L \setminus \{\varepsilon\}$$
$$U'_{n+1} = {U'_n}^{-1} L^\infty$$

With this sequence of sets, we have the following criterion to test strict-codes:

Theorem 2 [8] *Let $L \subset \Sigma^\infty$ be rational. L is a strict-code if and only if $U'_i = \varnothing$ for some $i \geq 1$.*

Example 5 *Let $L = a + (ab)^*ba$, then $U'_1 = (ba)^*bba$ and $U'_2 = \varnothing$ then L is a strict-code and an ω-code because L is a language of finite words.*

If the language L is a subset of Σ^* then Theorem 2 allows one to test ω-code without using infinite power of the language L because, in this case, the sequence of sets becomes:

$$U'_1 = L^{-1}L \setminus \{\varepsilon\}$$
$$U'_{n+1} = {U'_n}^{-1}L^*$$

As we will see in the section 5, the test can be implemented with an algorithm whose complexity in the worst case is $O(n^3)$ (n being the size of a deterministic automaton for L.)

4.3 Another test for Codes and ω-Codes

We introduce here a new algorithm also based on the computation of left quotients of languages, whose complexity in time is $O(n^3)$ too.
Let L be a subset of Σ^+ such that $L = root(L^*)$ and M denotes the monoid L^*. We define inductively the sequence $(V_n)_{n \geq 1}$ by:

$$V_1 = (M \setminus \{\varepsilon\})^{-1}L \setminus \{\varepsilon\}$$
$$V_{n+1} = (V_n M)^{-1}L$$

We can state the following results:

Result 1 *(theorem 3) A language L is a code if and only if none of the sets V_i contains the empty word.*

Example 6 *Let $L = \{b, abb, abbba, bbba, baabb\}$. Then we have $V_1 = \{bba, ba, a, aabb\}$, $V_2 = (V_1L^*)^{-1}L = \{abb, \varepsilon, bb, b, bbba, bba, ba, a\}$, $\varepsilon \in V_2$ and L is not a code.*

This result holds for every set of words (rational or not). Moreover for rational languages we can state:

Result 2 *(proposition 5) For a rational language L, all the sets V_i are rational languages and the number of the sets V_i is bounded by 2^n, n being the number of states of an automaton which recognizes L.*

This result shows that we can construct an algorithm based on the sequence $(V_n)_{n \geq 1}$, in $2^{card(\mathcal{A})}$ steps, to decide whether a given rational language is a code. With the following criterion, the same sequence allows one to decide if a rational language is an ω-code.

Result 3 *(theorem 4) A rational language L is an ω-code if and only if none of sets V_i is the empty set.*

Example 7 *Let $L = a + (ab)^*ba$,*

$$V_1 = (ba)^*bba \ , \quad V_2 = \varnothing \ \text{ then } \ L \text{ is an } \omega\text{-code.}$$

Before proving these results, let us state the following lemma.

Lemma 1 *For $i \geq 1$, V_i is the set of all i^{th} shifts for the factorizations of words of L^* over L.*

proof. We first prove by induction that each shift is an element of a set V_i. Let u and v be two words of L^+ comparable by the prefix order. Let f_u and f_v be two factorizations respectively for u and v, assuming that $f_{u1} \neq f_{v1}$. To each factorization f_u and f_v we associate respectively the sequence of left factors over L $(p_i)_1^n$ and $(q_i)_1^m$. Let us assume that $p_1 \lneq q_1$ and the sequence of shifts $s(f_u, f_v) = (s_1, \ldots, s_{max})$, where max is the number of shifts for the two factorizations f_u and f_v.
By definition, $s_1 \in (p_1L^*)^{-1}q_1 \subseteq V_1$. Let us assume that there exists $l, k > 1$ and $1 < j < max$ such that $s_j \in V_j$ and $s_j = (p_k)^{-1}q_l$ (or $s_j = (q_l)^{-1}p_k$). Consider that $s_j = (p_k)^{-1}q_l$, then by definition $s_{j+1} \in (s_jL^*)^{-1}L$ (see figure 2). Moreover, $s_j \in V_j$ then $s_{j+1} \in (s_jL^*)^{-1}L \subset (V_jL^*)^{-1}L = V_{j+1}$.

Figure 2. $s_j \in V_j$ then $s_{j+1} \in V_{j+1}$

Now, we prove by induction that each word in a set V_i is a shift. Let $u \in V_1$ then $u \in V1 \subseteq (L^+)^{-1}L$. Let $n \geq 1$ and for $u \in V_n$, this word verifies the lemma. Considering $v \in V_{n+1}$, there exists two words $u \in V_n$ and $u' \in L^*$

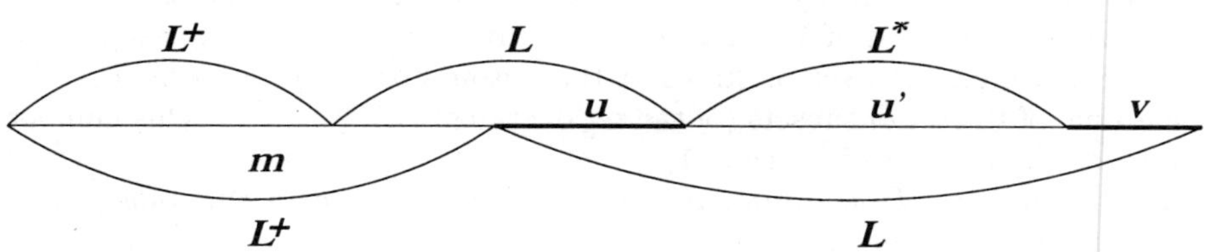

Figure 3. $v \in V_{n+1}$, $u \in V_n$ and $u' \in L^*$ then $uu'v \in L$

such that $uu'v \in L$ (see figure 3). As $u \in V_n$ there exists m such that for k

and l, $m = (p_i)_1^k$, $mu = (q_i)_1^l$, and $p_1 \neq q_1$ (inductive hypothesis), therefore $p_{k+1} = p_k uu'v = muu'v$ and for $l' > l$, $q_{l'} = q_l u' = muu'$. □

Theorem 3 *A language L over Σ is a code if and only if none of the sets V_i contains ε.*

proof. Let u be a word of L^+ which has two factorizations with different first factors. There exists $k \geq 1$, $l \geq 1$, and two sequences of left factors of $(p_i)_1^k$ and $(q_i)_1^l$ such that $p_k = q_l = u$ and $p_1 \neq q_1$. By lemma 1, there exists $r \geq 1$ such that $(p_k)^{-1} q_l \in V_r$, and then $\varepsilon \in V_r$.

By the lemma 1 with $u = \varepsilon \in V_i$ (for $i \geq 1$), there exists $m \in L^+$ with two sequences of left factors over L $(p_i)_1^k$ and $(q_i)_1^l$ and $p_1 \neq q_1$. So the word m of L^+ has two factorizations and therefore L is not a code if there exists $i \geq 1$ $\varepsilon \in V_i$. □

Lemma 2 *(corollary of lemma 1) If $\alpha \in L^\omega$ has two factorizations with different first factors over L, there exists $(p_i)_1^\infty$ and $(q_i)_1^\infty$ $(p_1 \neq q_1)$ and for every $n > 1$, there exists $l, k > 1$ such that $p_k < q_l < p_{k+1}$, such that $p_k^{-1} q_l \in V_n$.*

Proposition 4 *Let L be a language over Σ. If for every $i \geq 1$, $V_i \neq \varnothing$ then L is not an ω-code.*

proof. Suppose that L is not an ω-code. There exists an infinite word α which has two factorizations over L. There exists $(p_i)_1^\infty$ and $(q_i)_1^\infty$. By lemma 2, for $i \geq 1$, there exists $k \geq 1$ and $l \geq 1$ such that $p_k^{-1} q_l \in V_i$ and then V_i is not empty. □

Rational case

Our object is to construct an algorithm to decide if a rational language is an ω-code. In order to do so, we must be able to decide when we can stop the construction of the sequence $(V_i)_{i \geq 1}$. The next result establish the fact that the number of sets V_i is finite for a given rational language.

Proposition 5 *Let L be a rational language over Σ. The cardinal of the set of V_i for L is bounded by $2^{card(\mathcal{A}(L))}$ ($\mathcal{A}(L)$ is an automaton for L).*

proof. A word of V_n $(n \geq 1)$ is a suffix of a word of L. A set V_n can be represented by a set of states of the automaton $\mathcal{A}$ recognizing L. The cardinal of the set of subsets (of the set of states) is $2^{card(\mathcal{A}(L))}$. This number bounds the number of V_n $(n \geq 1)$. □

Theorem 4 *Let L be a rational language. L is an ω-code if and only if there exists $i \geq 1$ such that $V_i = \varnothing$.*

proof. By the proposition 4 if there exists $i \geq 1$ such that $V_i = \varnothing$ then L is an ω-code.

Conversely, let us suppose that for $i \geq 1$, $V_i \neq \varnothing$ and set $n = card(\mathcal{A}(L))$. Let $v \in V_{m_0}$, $m_0 > n$ then, by lemma 1, there exist $p = (p_i)_1^{j_1}$ and $q =$

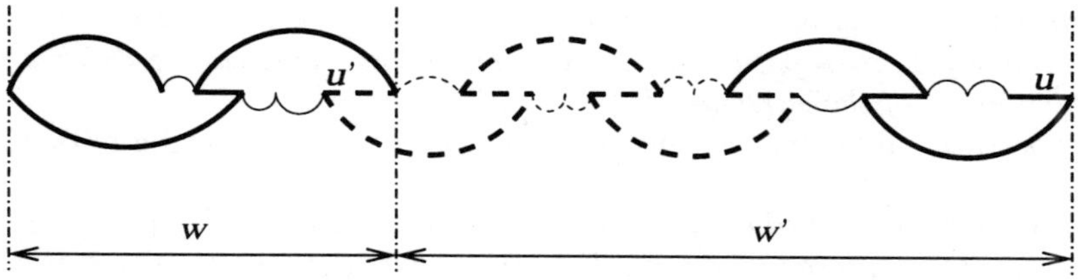

Figure 4. $u \sim_L u'$

$(q_i)_1^{j_2}$ such that $v = p_{j_1}^{-1} q_{j_2}$. Let us denote by $s(p,q)$ the sequence of shifts associated to the two factorizations induced by the sequences p and q of left factors. There exists m_1 and m_2, $1 \leq m_1 \neq m_2 \leq m_0$ such that $u = s_{m_1}(p,q)$ and $u' = s_{m_2}(p,q)$, with $u \sim_L u'$ (see figure 4.) For u there exists k and l such that $u = p_k^{-1} q_l$ (or $u = q_l^{-1} p_k$), and for u' there exists r and t such that $u' = p_r^{-1} q_t$ (or $u' = q_t^{-1} p_r$).
Let $w = p_r$ if q_t is prefix of p_r, (else let $w = q_t$). Let $w' = w^{-1} q_l$. Let $u'' \in pref(w')$ such that $u'u'' \in L$ and let us consider the two following cases :

- $u' = p_r^{-1} q_t$, then $w' \in L^+$ and $((u'')^{-1}) w' (u^{-1}) \in L^+$, $u'u'' \in L$, and $uu'' \in L$ $(u \sim_L u')$.

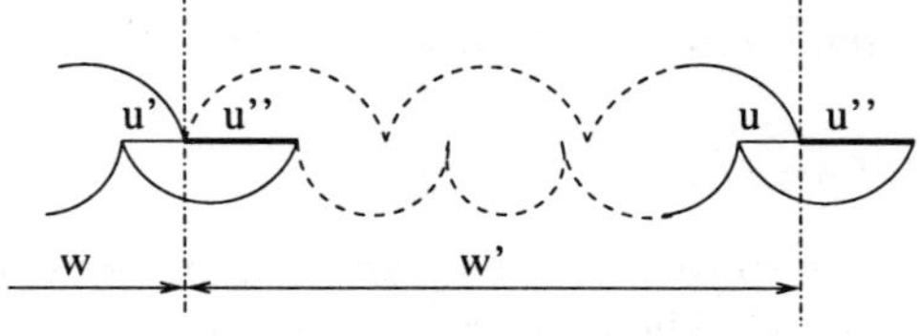

- $u' = q_t^{-1} p_r$, then $w'(u^{-1}) \in L^+$ and $((u'')^{-1}) w' \in L^+$, $u'u'' \in L$, and $uu'' \in L$ $(u \sim_L u')$.

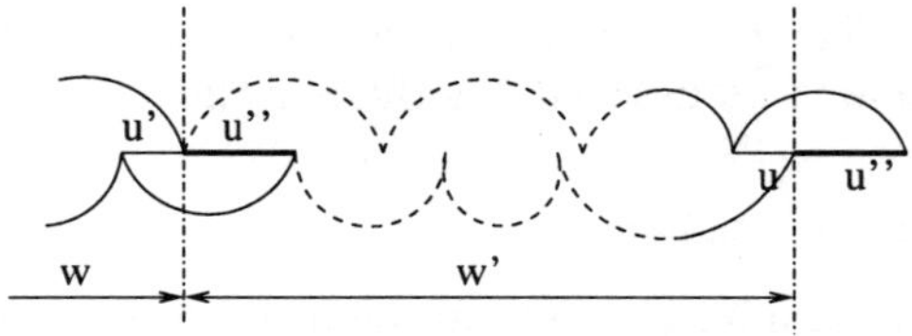

The infinite word $\alpha = w(w')^\omega$ has two factorizations over L. □

Remark 5 *If L is not a code then it is not an ω-code, and then, if $\varepsilon \in V_i$ for some $i \geq 1$, there is no V_j that are empty sets. This can be checked with the definition of the sequence $(V_i)_{i \geq 1}$.*
For non-rational languages, this theorem does not hold (see the next example).
Example 8 *Let L be the following language $a^+b \cup \{a^nba^{n-1}c, a^nca^{n-1}c/n \geq 1\}$. L is not a rational language. L is an ω-code, but for $n \geq 1$, $c \in V_n$.*
On the figure 5, we can see that for $n \geq 1$, $a^{n-1}c \in V_1$, $a^{n-2}c \in V_2$, ..., $ac \in V_{n-1}$, and $c \in V_n$.

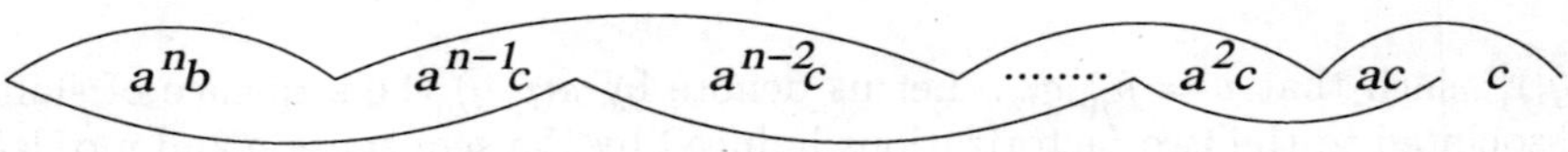

Figure 5. $\forall n \geq 1$, $c \in V_n$

5 Complexity analysis

The complexity analysis of algorithms to test finite codes can be found in [2],[3] and [4] (in $O(n.m)$, n being the number of words in the language and m being the sum of the length of words). For a rational language given by a deterministic automaton of n states, ones can decide whether the language is a code and the complexity is $O(n^2)$[6].
For the tests presented in section 4, we give the followings complexities:

Test by interlaced product

Let L be a language to test, let $\mathcal{A}$ be an unambiguous normalized finite automaton recognizing L. The main part of the complexity of this test consists in deciding if there is a cycle in the automaton $\mathcal{A}'$ which contains at least two states of type (q_0, q) and (q', q_0) (for q and q' different of q_0). This can be done in $O(n^3)$. In effect, there is at most n states of type (q_0, q) in $\mathcal{A}'$. For each of this state, ones has to search states of type (q', q_0) in the set of its descendents. There are also at most n such states. For each (q', q_0) one must check if among its descendents there is the state (q_0, q).

Tests by quotients of languages

Let $\mathcal{A}(L)$ be a deterministic finite automaton recognizing L. The two algorithms of sections 4.2 and 4.3 do the same kinds of operations and the same number of times.

The second part of the proof of the theorem 4 tell us that ones needs to compute at most $n = card(\mathcal{A})$ sets $V_{i \geq 0}$ to decide if a rational language is an ω-code:

Lemma 3 *If there exists $i \geq 1$ such that $V_i = \emptyset$ then there exists $j \leq n+1$ such that $V_j = \emptyset$.*

algorithm of section 4.2

[1]- $M = L^*$

[2]- $U_1 = (L^{-1}L) \setminus \{\varepsilon\}$

[3]- For $i = 2$ to $n+1$ do

- **[3a]**- $U_i = U_{i-1}^{-1}M$
- **[3b]**- if $U_i = \varnothing$ then L is an ω-code.
- **[3c]**- if $\varepsilon \in U_i$ then L is not a code.

[4]- if $U_n \neq \varnothing$ then L is not an ω-code.

algorithm of section 4.3

[1]- $M = L^*$

[2]- $V_1 = ((M \setminus \{\varepsilon\})^{-1}L) \setminus \{\varepsilon\}$

[3]- For $i = 2$ to $n+1$ do

- **[3a]**- $V_i = (V_{i-1}M)^{-1}L$
- **[3b]**- if $V_i = \varnothing$ then L is an ω-code.
- **[3c]**- if $\varepsilon \in V_i$ then L is not a code.

[4]- if $V_n \neq \varnothing$ then L is not an ω-code.

The main part of the complexity for this algorithm is done by the step **[3a]**, the left quotient $(V_{i-1}M)^{-1}L$. The algorithm given in [14] compute the left quotient of two non deterministic finite automata has a complexity in $O(n^3)$ if one automaton is deterministic, that is the case for our construction. The number of V_i that we have to compute is bounded by n, because the number of states of $\mathcal{A}(L)$ is n. The complexity in time of this algorithm is $O(n^4)$.
For each state q of $\mathcal{A}(L) = (Q, I, F, \delta, \Sigma)$, L_q is the language recognizing by the automaton $\mathcal{A}_q = (Q, q, F, \delta, \Sigma)$ (like $\mathcal{A}(L)$, this automaton is deterministic). To improve efficiency of the algorithm, we can precompute, for each state $q \in Q$, the set of state V_q defined by the set of the initials states of the automaton of $(L_qM)^{-1}L$, the time to execute this step is $O(n^2)$ with

the algorithm for left quotient given in [14] because $\mathcal{A}_q$ is deterministic. The complexity of this preprocessing is $O(n^3)$. The step **[3a]** ($V_i = (V_{i-1}M)^{-1}L$) become $V_i = \bigcup_{q \in V_{i-1}} V_q$, and then the complexity in time for the algorithm is $O(n^3)$.

Acknowledgments

The authors would like to thank the anonymous referee for his careful reading and suggestions.

References

1. L. Staiger. On infinitary finite length codes. *Theoretical Informatics and Applications*, 20(4):483–494, 1986.
2. S. Even. *Graph algorithms.* Addison-Wesley,Reading,MA, 1973.
3. M. Rodeh. A fast test for unique decipherability based on suffix trees. *IEEE transaction on information theory*, 28:648–651, 1982.
4. A. Apostolico and R. Giancarlo. Pattern matching machine implementation of a fast test for unique decipherability. *Inform. Proc. Letters*, 18:155–158, 1984.
5. A.A. Sardinas and C.W. Patterson. A necessary and sufficent condition for the unique decomposition of coded messages. *IRE Internat. Conv. Rec.*, 8:104–108, 1953.
6. J. Berstel and D. Perrin. Trends in the theory of codes. *Bull. of EATCS*, 29:84–95, 1986.
7. R. König. Lectures on codes. 1994.
8. Nguyen Huong Lâm and Do Long Van. On a class of infinitary codes. *Theoretical Informatics and Applications*, 24(5):441–458, 1990.
9. J.E. Hopcroft and J.D. Ullman. *Introduction to automata theory, languages, and computation.* Addison-Wesley, 1979.
10. D. Perrin and J.E. Pin. Infinite words. (to be published) http://www.liafa.jussieu.fr/~jep/Resumes/InfiniteWords.html.
11. J. Karhumäki. A property of three-element codes. *Theoret. Comput. Sci.*, 41:215–222, 1985.
12. J. Berstel and D. Perrin. *Theory of codes.* Academic Press, 1985.
13. Do Long Van. *Contribution to Combinatorics on Words.* Doctor B thesis, Humboldt University, Berlin, 1985.
14. O. Matz, A. Miller, A. Potthoff, W. Thomas, and E. Valkema. Report on the program AMoRE. Technical report, Institut für Informatik und Praktische Mathematik, 1995. Software.

Distributed Random Walks for an Efficient Design of a Random Spanning Tree

Hichem Baala   Marc Bui †

Abstract

We present a distributed algorithm for constructing a random spanning tree, making use of random walks as network traversal scheme. Our approach is novel and make use of distributed random walks, each one represented by a token annexing a territory over the underlying graph. These multiple random walks collapse into a final one, that defines the final territory and provides the random spanning tree. The scheme is parallel and make use of waves to merge very efficiently the spanning forest computed by the random walks into one final random spanning tree.
Keywords: random spanning tree, random walks, distributed algorithm.

1 Introduction

A *distributed algorithm* is an algorithm designed to run on a distributed system where many processes cooperate to solve parts of a given problem in parallel. The problem of efficiently constructing a spanning tree in distributed networks is a central one and is essential for structuring a distributed system. We address the problem of constructing such a structure with a protocol that tolerate faults and adapt itself to dynamic topology changes. In this paper, we introduce Distributed Random Walks (DRW) as a collection of random walks that cooperate in order to establish a computation. The technique uses a collection of random walks that are coalescing into a final one which maintains the control structure. We apply this technique to compute a spanning tree which is randomly selected among all the possible ones for the network, and to gather informations, we use a wave scheme. We can informally described the whole procedure as follows : several nodes initiate a random walk, with an *explorer token.* Every node, upon receiving an explorer token, mark himself visited with the identity of the token, except if it has already been visited by another token, and then forwards at random to one of its neighbors the received explorer token. The network is thus, explored in parallel and decomposed into subregions, one per token. Each token constructs a subtree of the network. When a node meet another one, or an already visited node, a wave is initiated. This wave is a backward propagation wave that merges one of the subtrees with the other into one. This process is driven in parallel and eventually, the waves will cover the network, resulting in the spanning tree definition and the protocol is ready for termination when a single explorer token remains and all nodes of the graph are visited. In this paper, we develop a technique for designing algorithms on graphs, especially for an efficient random spanning tree. Our goal is the computation of a random spanning tree (ie a spanning tree that is chosen randomly among all possible spanning trees).

The problem. Let $G(V, E)$ be a connected graph representing a distributed system with real-valued weights $w : E \quad R$ having n vertices and m edges. A spanning tree in G is an acyclic subgraph of G that includes every vertex of G and is connected; every spanning tree has exactly $n - 1$ edges. We are interested in computing in a distributed way a random spanning tree (RST) (i.e. a random spanning tree is a spanning tree selected among all possible spanning tree on the underlying graph at random).

Université de Reims, Département de Mathématiques et Informatique, UFR Sciences Exactes et Naturelles, Moulin de la Housse, BP 1039, F-51687 Reims Email. Hichem.Baala@univ-reims.fr

†LRIA, Université Paris 8, 15 rue Catulienne, 93200 Saint-Denis, France, Email. Marc.Bui@univ-paris8.fr

Related Works. In distributed computing, the design of a spanning tree structures for distributed networks has a wide litterature. Many papers [GHS83], [LR86], [GKP93] propose distributed solutions and identify interesting properties to construct efficiently and in parallel a spanning tree. Else, the power of RW has also been demonstrated in distributed computing, several authors have successfully designed original solutions for many important control problems such as mutual exclusion [IJ90] or Unique Naming problem [AE91], within the self-stabilizing area where the goal is to cope with possible transient failures. We will make use of the attractive techniques found in this area in our solution. E. Chang [Chang82] and A. Segall [Segall83] have introduced the concept of *wave* in distributed computing with the *Propagation of Information with Feedback* scheme. More recently, [FGL94] and [LB99] have respectively defined the distributed recursive wave and distributed recursive multiwave as a general programming paradigms for distributed systems. We couple these items with a derivation of spanning tree construction and RW techniques to define our general modular technique that we call DRW.

Contributions In this paper, we propose a multiple random walks scheme combined with a generalized diffusing feedback scheme (waves) that allow a fast RST construction. The main result of this paper, is that the simulation of multiple random walks on a connected undirected graph G coupled with some (adequate) path reversal scheme (that we call *waves*) can be used to generate a spanning tree of G at random. The advantages of our scheme are the following: we generate a random spanning tree structure that is less subject to failures compare to a deterministically predetermined spanning tree; the solution is adaptive and deals with topology changes and can be adapted to ad-hoc wireless networks; it can be derived to obtain a self-stabilizing solution; it is parallel, uses the whole power of distributed ressources and exhibits a good average running time.

Outline of the Paper. The paper is organized as follows : in the next section, we describe the model of distributed computation assumed. In section 3, we describe the algorithm illustrating the use of multiple random walks for selecting a random spanning tree. Thereafter, we show how we implement thisand discuss why it is an efficient solution and discuss the key points on creating a RST. Section 4 addresses the (time and message) complexity aspects as well as some remarks based on simulations . An informal correctness proof is also given in this section. Finally, in section 5, we give some concluding remarks and open questions.

2 Preliminaries

In this section, we give the definitions needed and we introduce some of the tools we use.

The system. We model the network as an undirected connected graph $G = (V, E)$ with V the set of nodes ($V = n$) and E s the set of edges. Each node represents a computer and each link represents a bidirectional communication channel. Each node is associated to a unique identifier A communication link (i, j) exists iff i and j are neighbors, and is asociated to a cost which can vary in time but is always positive. A change in the status of a node is implicitly recognized by the change in the status of its links. We consider the network to be asynchronous. Each node i maintains its set of neighbors, denoted as N_i. The degree of i is the number of neighbors of i, it is equal to N_i .

Processes. Every process of the distributed systems executes the same code. The program consists of a set of *variables* and a finite set of *rules*. A process proceed to an internal action (for example, write to its own variables, compute something or send a message) upon reception of a message.

Random Walk. Let us consider a token that moves on a connected undirected graph $G = (V, E)$. At each step, the token goes from the current vertex to one of its neighbors, chosen uniformly at random. This stochastic process is a Markov chain; it is called (simple) random walk on the graph.

3 Algorithms description

Data Structure Each node p maintains
- $color$, the identity of a token
- $master$, the (sub)tree root which the node belongs to
- $father$,the node father within the (sub)tree
- $sons$ set of sons ,is the set of the node sons

The algorithm RST is specified in a pseudocode form as in [KKM90] for a better understanding.We specify the algorithm behavior by means of overall actions driven by tokens and waves.

Some sites randomly generate token identified by a color and charaterized by the initiators of the token.

(TA) Token Annexing Mode whenever a $token_i(color_i, rac_i)$ issued from a node q is annexing (or generated at) node p, which belongs to a (sub)tree [i.e. a $token_j(color_j, rac_j)$]

(TA1) if $color_i < color_j$, the annexing is stopped and the token is destroyed.

(TA2) if $color_i > color_j$, one of the 2 conditions holds :

- (i) one (or more) token(s) are present on node p
- (ii) no other token on node p

(TA2-i) if test collision is true :

if $token_i$ is the unique biggest, it continue its traversal and all others are destroyed. Node p marked himself with $color$ $color_i$, $master$ rac_i, $father$ q

if $token_i$ is biggest but not unique (others $token_{j_1}, \ldots, token_{j_d}$ has respectively $color_{j_1}, \ldots, color_{j_d}$ equal to $color_i$, $token_i$ and $token_{j_1}, \ldots, token_{j_d}$ are merged to form the unique token of identity $i+1$ rooted in p [i.e. $token(i+1, p)$ is generated]

if $token_i$ is not the biggest, it is destroyed.

(TA2-ii) $token_i$ continue its traversal scheme. p marked himself with $color$ $color_i$, $master$ rac_i, $father$ q

(WU) Wave Update Mode Whenever a $token_i(color_i, rac_i)$ reaches a node p with its variable $color$ such that $color < color_i$ a wave is generated.

WU1 the wave is propagated applying a path reversal scheme over the domain identified by $color$ (the domain which p belongs to)

WU2 the wave stops itself when it reaches the p domain limit.

Termination of the algorithm is realized with a derivation of the Dijkstra-Scholten scheme [DS80] known as diffusing computation. This termination detection is periodically initiated by nodes which have initiated an annexing token.

Example The following example illustrates RST's construction.

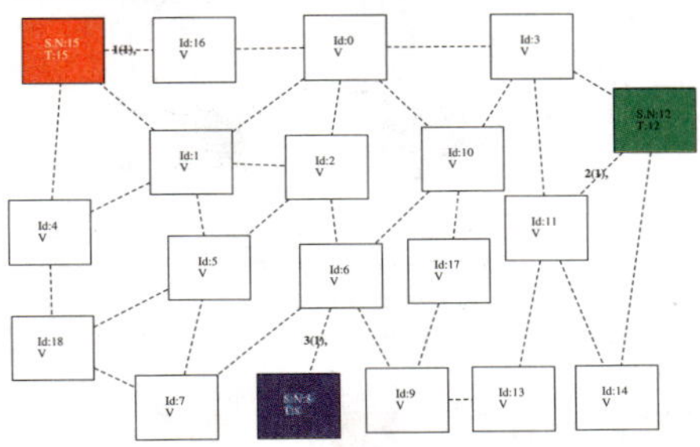

Figure 1: Example network. 3 token proceed to random walks.

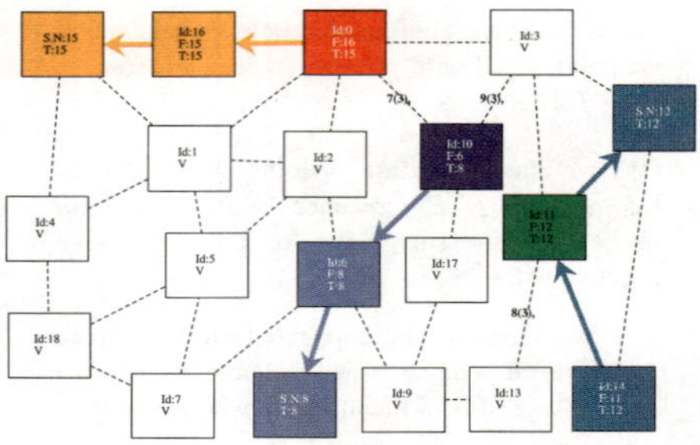

Figure 2: Evolution of random walks: 3 explorer tokens annexing regions of the graph in parallel, blue red and green (the tokens are located in darker color)

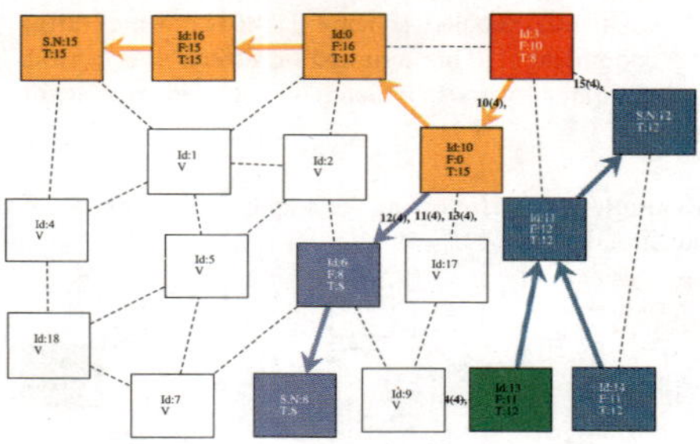

Figure 3: Waves: a wave is initiate when a token meets another one or enter the region of the graph already explored by a token. In this example, the meeting of the red and blue token initiate a red wave (that will flood the blue region

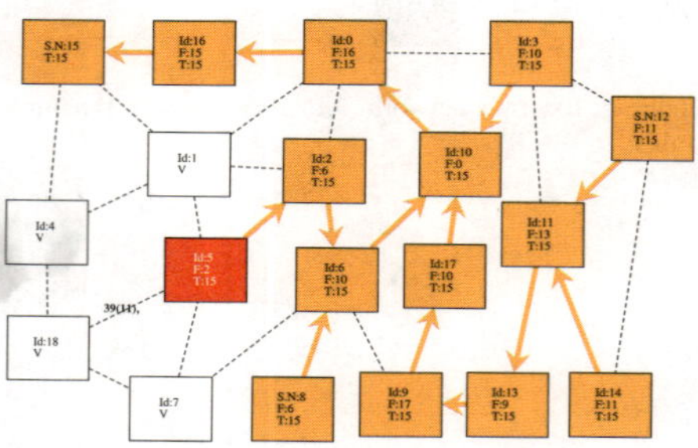

Figure 4: Evolution of waves: the wave issued from the token with the largest identity merge the subtrees and update the variables of each node (here the red token wins).

4 Correctness and complexity proofs

The proof is by the following lemmas. The proofs of the lemmas are not detailed in this extended abstract.

Lemma 1 *At least a token starts a random walk.*

Lemma 2 *Within a finite time k tokens will eventually collide and only one token remains after.*

Lemma 3 *After a finite time, a single token is in the network, it will eventually visits all the nodes*

Lemma 4 *A wave is eventually initiated and will update all visited nodes.*

Lemma 5 *The algorithm terminates and a spanning tree is ouput.*

Theorem 1 *The algorithm outputs a random spanning tree.*

We have tested the algorithm with a similator written in C++ with the LEDA library. Under the assumption ofa synchronous distributed system, the results obtained show a good time complexity.

Network size (n,m)	initiators $\sqrt{n}$ average	Time (steps) average	Nb msg
(30,46)	5	52	242
(40,65)	6	66	344
(50,96)	8	73	460

5 Conclusions and future works

We have presented DRW an efficient scheme for constructing a uniform spanning tree over a distributed network. It is based on a new family of algorithms for distributed computing called the distributed random walks scheme algorithms. The key advantage of this novel approach is that it has a very moderate and the best balanced impact on network and computer resources. RW are expected to converge to one that keeps a surveillance function to relaunch computation in case of links or node failures.

However, the estimate of the complexity of DRW algorithms for the general case is more complicated than for usual DC algorithms The difficulty is due to the interactions between the random walks.

References

[ACK+98] B. Awerbuch, I. Cidon, S. Kutten, Y. Mansour, and D. Peleg. Optimal broadcast with partial knowledge. *Siam J. Comput*, 28(2):511–524, 1998.

[AEY91] E. Anagnostou and R. El-Yaniv. More on the power of random walks: Uniform randomized algorithms. In *Proceedings of the 5th International Wokshop on Distributed Algorithms and Graphs*, 1991.

[AKL+79] R. Aleliunas, R. Karp, R. Lipton, L. Lovasz, and C. Rackoff. Random walks universal traversal sequences and the complexity of maze problems. *IEEE*, pages 218–222, 1979.

[Ald90] David J. Aldous. The random walk construction of uniform spanning trees and uniform labelled trees. *Siam J. Disc. Math.*, 3(4):450–465, november 1990.

[Awe85] Baruch Awerbuch. A new distributed depth-first-search algorithm. *Information Processing Letters*, 20:147–150, april 1985.

[BFSU98] A. Broder, A. Frieze, S. Suen, and E. Upfal. Optimal construction of edge-disjoint paths in random graphs. *Siam J. Comput*, 28(2):541–573, 1998.

[BIZ89] J. Bar-Ilan and D. Zernik. Random leaders and random spanning trees. In JC. Bermond M. Raynal, editor, *International Workshop on Distributed Algorithms and Graphs*, Lecture Notes in Computer Science 392, pages 1–12, Nice, France, September 1989. Springer.

[BK89] A. Z. Broder and A. R. Karlin. Bounds on the cover time. *Journal of Theoretical Probability*, 2(1):101–120, 1989.

[Cha92] E. Chang. Echo algorithms: Depth parallel operations on general graphs. *IEEE Transactions on Software Engineering*, 8(4):391–401, 192.

[DMP82] D. Dolev, J. Meseguer, and M.C. Pease. Finding safe paths in a faulty environment. *ACM Symposium On principles Of Distributed Computing*, pages 95–103, 1982.

[DS80] E.W. Dijkstra and C.S. Scholten. Termination detection for diffusing computations. *Information Processing Letters*, 11(1):1–4, 1980.

[Epp98] D. Eppstein. Finding the k shortest paths. *Siam J. comput*, 28(2):652–673, 1998.

[FFK+98] A. Fiat, D. Foster, H. Karloff, Y. Rabani, Y. Ravid, and S. Vishwanathan. Competitive algorithms for layered graph traversal. *Siam J. Comput*, 28(2):447–462, 1998.

[GHS83] Robert G. Gallager, Pierre A. Humblet, and Paul M. Spira. A distributed algorithm for minimum weight spanning trees. *TOPLAS*, 5(1):66–77, 1983.

[GKP93] Juan A. Garay, Shay Kutten, and David Peleg. A sub-linear time distributed algorithm for minimun-weight spanning trees. *IEEE*, pages 659–667, 1993.

[IJ90] A. Israeli and M. Jalfon. Token management schemes ans random walks yields self-stabilizing mutual exclusion. In *Proceeedins of the 9th ACM Symposium on Principles of Distribbuted Computing*, 1990.

[KKM90] E. Korach, S. Kutten, and S. Moran. A modular technique for the design of efficient distributed leader finding algorithms. *ACM Transactions on Programming Languages and Systems*, 12(1):84–101, 1990.

[KLNS89] J.D. Kahn, N. Linial, N. Nisan, and M.E. Saks. On the cover time of random walks on graphs. *Journal of Theoretical Probability*, 2(1):121–128, 1989.

[KT94] P. Klein and R. E. Tarjan. A randomized linear-time algorithm for finding minimum spanning trees. *ACM*, pages 9–15, 1994.

[PW98] James G. Propp and David B. Wilson. How to get an exact sample from a generic markov chain and sample a random spanning tree from a directed graph, bith within the cover time. *Journal of Algorithms*, pages 170–217, 1998. Combines two conference article, one appearing in 1996 ACM-SIAM Symposium on Discrete Algorithm, the second appearing in ACM Symposium on the Theory of Computing.

[Seg83] A. Segall. Distributed network protocols. *IEEE Transactions on Information Theory*, 29(1):23–35, 1983.

[STU97] Akiyoshi Shioura, Akihisa Tamura, and Takeaki Uno. An optimal algorithm for scanning all spanning trees of undirected graphs. *Siam J. Comput*, 26(3):678–692, 1997.

[TW91] P. Tetali and P. Winkler. On an random walk problem arising in self-stabilizing token management. In *Proceedings of the Tenth Annual ACM Symposium on Principles of Distributed Computing*, pages 273–280, 1991.

[WG94] J. Warpechowska-Gruca. Meeting times of random walks on graphs. Technical Report 94/535/04, University of Calgary, 1994.

Formal Concept Analysis and Rough Set Theory in Clustering

Ho Tu Bao
Japan Advanced Institute of Science and Technology, Japan
National Institute of Information Technology, Vietnam

Abstract. This paper is concerned with the fundamental role of two mathematical theories in some clustering problems. Formal concept analysis provides the algebraic structure and properties of possible concepts from a given context, and rough set theory provides a mathematical tool to deal with imprecise and incomplete data. Based on these theories, we developed models and algorithms for solving three clustering problems: conceptual clustering, approximate conceptual clustering, and text clustering.

1 Formal Concept Analysis and Rough Set Theory

A theory of concept lattices has been studied under the name *formal concept analysis* (FCA) by Wille and his colleagues [1, 11]. Considers a *context* as a triple $(\mathcal{O}, \mathcal{D}, \mathcal{R})$ where $\mathcal{O}$ be a set of objects, $\mathcal{D}$ be a set of primitive descriptors and $\mathcal{R}$ be a binary relation between $\mathcal{O}$ and $\mathcal{D}$, i.e., $\mathcal{R} \subseteq \mathcal{O} \times \mathcal{D}$ and $(o, d) \in \mathcal{R}$ is understood as the fact that object o has the descriptor d. For any object subset $X \subseteq \mathcal{O}$, the largest tuple common to all objects in X is denoted by $\lambda(X)$. For any tuple $S \in \mathcal{T}$, the set of all objects satisfying S is denoted by $\rho(S)$. A tuple S is *closed* if $\lambda(\rho(S)) = S$. Formally, a *concept* C in the classical view is a pair (X, S), $X \subseteq \mathcal{O}$ and $S \subseteq \mathcal{T}$, satisfying $\rho(S) = X$ and $\lambda(X) = S$. X and S are called *extent* and *intent* of C, respectively. Concept (X_2, S_2) is a *subconcept* of concept (X_1, S_1) if $X_2 \subseteq X_1$ which is equivalent to $S_2 \supseteq S_1$, and (X_1, S_1) is then a *superconcept* of (X_2, S_2).

It was shown that λ and ρ define a Galois connection between the power sets $\wp(\mathcal{O})$ and $\wp(\mathcal{D})$, i.e., they are two order-reversing one-to-one operators. As a consequence, the following properties hold which will be exploited in the learning process:

$$\begin{array}{c}
\text{if} \quad S_1 \subseteq S_2 \quad \text{then} \quad \rho(S_1) \supseteq \rho(S_2) \quad \text{and} \quad \lambda\rho(S_1) \subseteq \lambda\rho(S_2) \\
\text{if} \quad X_1 \subseteq X_2 \quad \text{then} \quad \lambda(X_1) \supseteq \lambda(X_2) \quad \text{and} \quad \rho\lambda(X_1) \subseteq \rho\lambda(X_2) \\
S \subseteq \lambda\rho(S), \quad X \subseteq \rho\lambda(X) \\
\rho\lambda\rho = \rho, \quad \lambda\rho\lambda = \lambda, \quad \lambda\rho(\lambda\rho(S)) = \lambda\rho(S) \\
\rho(\bigcup_j S_j) = \bigcap_j \rho(S_j), \quad \lambda(\bigcup_j X_j) = \bigcap_j \lambda(X_j)
\end{array}$$

The basic theorem in formal concept analysis [11] states that the set of all possible concepts from a context $(\mathcal{O}, \mathcal{D}, \mathcal{R})$ is a *complete lattice*[1] $\mathcal{L}$, called Galois lattice, in which infimum and supremum can be described as follows:

$$\bigwedge_{t\in T} (X_t, S_t) = (\bigcap_{t\in T} X_t, \lambda\rho(\bigcup_{t\in T} S_t)) \tag{1}$$

$$\bigvee_{t\in T} (X_t, S_t) = (\rho\lambda(\bigcup_{t\in T} X_t), \bigcap_{t\in T} S_t) \tag{2}$$

Rough set theory, a mathematical tool to deal with uncertainty introduced by Pawlak in early 1980s [10]. The starting point of this theory is the assumption that our "view" on elements of a set of objects $\mathcal{O}$ depends on some equivalence relation E on $\mathcal{O}$. An *approximation space* is a pair $(\mathcal{O}, E)$ consisting of $\mathcal{O}$ and an equivalence relation $E \subseteq \mathcal{O} \times \mathcal{O}$.

The key notion of the rough set theory is the *lower* and *upper approximations* of any subset $X \subseteq \mathcal{O}$ which consist of all objects *surely* and *possibly* belonging to X, respectively. The lower approximation $E_*(X)$ and the upper approximation $E^*(X)$ are defined by

$$E_*(X) = \{o \in \mathcal{O} : [o]_E \subseteq X\} \tag{3}$$

$$E^*(X) = \{o \in \mathcal{O} : [o]_E \cap X \neq \emptyset\} \tag{4}$$

where $[o]_E$ denotes the equivalence class of objects indiscernible with o with respect to the equivalence relation E.

2 FCA-based Conceptual Clustering

Conceptual clustering concerns mainly with symbolic data [9]. It does simultaneously two tasks: (i) *hierarchical clustering* (i.e., finding a hierarchy of useful subsets of unlabelled instances); and (ii) *characterization* (i.e., finding an intensional definition for each of these instance subsets). An important feature of conceptual clustering is that a partitioning of data is viewed as

[1] A lattice $\mathcal{L}$ is complete when each of its subsetf X has a least upper bound and a greatest lower bound in $\mathcal{L}$.

Table 1: Scheme of OSHAM conceptual clustering

Input concept hierarchy H and an existing splittable concept C_k.
Result H formed gradually.
Top-level call OSHAM(root concept, $\emptyset$).

1. While C_k is still splittable, find a new subconcept of it that corresponds to the hypothesis minimizing the quality function $q(C_k)$ among η hypotheses generated by the following steps

   (a) Find a "good" attribute-value pair concerning the best cover of C_k.

   (b) Find a closed attribute-value subset S containing this attribute-value pair.

   (c) Form a subconcept C_{k_i} with the intent is S.

   (d) Evaluate the quality function with the new hypothesized subconcept.

   Form intersecting concepts corresponding to intersections of the extent of the new concept with the extent of existing concepts excluding its superconcepts.

2. If one of the following conditions holds then C_k is considered as unsplittable

   (a) There exist not any closed proper feature subset.

   (b) The local instances set C_k^r is too small.

   (c) The local instances set C_k^r is homogeneous enough.

3. Apply recursively the procedure to concepts generated in step 1.

'good' if and only if each cluster has a 'good' conceptual interpretation. In this sense, FCA is a good tool for conceptual clustering as it formalizes the duality between objects and their properties by Galois connections. Based on FCA, we have developed a conceptual clustering method OSHAM with some additional components to the concept representation by extent and intent. The key idea here to enrich the concept representation in FCA by adding several components based on the probabilistic and exemplar views on concepts that allow dealing better with typical or unclear cases in the region boundaries. The conceptual clustering method OSHAM to form a concept hierarchy in the framework of concept lattices is introduced and described in [2]. OSHAM searches to extract a good concept hierarchy by exploiting the structure of Galois lattice of concepts as the hypothesis space.

Instead of characterizing a concept only by its intent and extent, OSHAM represents each concept C_k in a concept hierarchy $\mathcal{H}$ by a 10-tuple

$$< l(C_k), f(C_k), s(C_k), i(C_k), e(C_k), d(C_k), p(C_k), d(C_k^r), p(C_k^r|C_k), q(C_k) > \tag{5}$$

where

- $l(C_k)$ is the level of C_k in $\mathcal{H}$;
- $f(C_k)$ is the list of direct superconcepts of C_k;
- $s(C_k)$ is the list of direct subconcepts of C_k;
- $i(C_k)$ is the intent of C_k (set of all common properties of instances of C_k);
- $e(C_k)$ is the extent of C_k (set of all instances satisfying properties of $i(C_k)$);
- $d(C_k)$ is the dispersion between instances of C_k;
- $p(C_k)$ is the occurrence probability of C_k;
- $d(C_k^r)$ is the dispersion of local instances of C_k which are not classified into subconcepts of C_k;
- $p(C_k^r|C_k)$ is the conditional probability of these unclassified instances of C_k;
- $q(C_k)$ is the quality estimation of splitting C_k into subconcepts C_{k_i}.

Table 1 represents the essential idea of algorithm OSHAM that allows discovering both disjoint and overlapping concepts depending on the user's interests by refining the condition 1.(a) and the intersection operation. In short, OSHAM combines the concept intent, hierarchical structure information, probabilistic estimations and the nearest neighbors of unknown instances. A experimental comparative evaluation of OSHAM is given in [2].

3 Approximate Conceptual Clustering

Kent[7] has pointed out common features between formal concept analysis and rough set theory, and formulated the *rough concept analysis* (RCA). For the sake of simplicity, we restrict ourselves here to present the basic idea of presenting approximate concepts in case of binary attributes where $\mathcal{D}$ is identical to the set $\mathcal{A}$ of all attributes a. Saying that a given formal context $(\mathcal{O}, \mathcal{A}, \mathcal{R})$ is not obtained completely and precisely means that the relation $\mathcal{R}$ is incomplete and imprecise. Let $(\mathcal{O}, E)$ be any approximation space on objects $\mathcal{O}$, we wish to approximate $\mathcal{R}$ in terms of E. The lower approximation $\mathcal{R}_{*E}$ and the upper approximation $\mathcal{R}^{*E}$ of $\mathcal{R}$ w.r.t. E can be defined element-wise as

$$\mathcal{R}_{*E}a = E_*(\mathcal{R}a) = \{o \in \mathcal{O} \mid [o]_E \subseteq \mathcal{R}a\} \tag{6}$$

$$\mathcal{R}^{*E}a = E^*(\mathcal{R}a) = \{o \in \mathcal{O} \mid [o]_E \cap \mathcal{R}a \neq \emptyset\} \tag{7}$$

The formal context $(\mathcal{O}, \mathcal{A}, \mathcal{R})$ can be then roughly approximated by two lower and upper formal contexts $(\mathcal{O}, \mathcal{A}, \mathcal{R}_{*E})$ and $(\mathcal{O}, \mathcal{A}, \mathcal{R}^{*E})$. These approximate contexts can be intuitively viewed as "truncated" and "filled up" contexts with respect to the equivalence relation E. Two formal context $(\mathcal{O}, \mathcal{A}, \mathcal{R})$ and $(\mathcal{O}, \mathcal{A}, \mathcal{R}')$ are *E-roughly equal* if they have the same lower and upper formal contexts, i.e., $\mathcal{R}_{*E} \equiv \mathcal{R}'_{*E}$ and $\mathcal{R}^{*E} \equiv \mathcal{R}'^{*E}$. A *rough formal context* in $(\mathcal{O}, E)$ is a collection of formal contexts of object set $\mathcal{O}$ and attribute set $\mathcal{A}$ which have the same lower and upper formal contexts (roughly equal formal contexts).

The rough extent of an attribute subset $S \subseteq \mathcal{A}$ w.r.t. $\mathcal{R}_{*E}$ and $\mathcal{R}^{*E}$ are defined as

$$\rho(S_{*E}) = \bigcap_{a \in S} \mathcal{R}_{*E}a \qquad \rho(S^{*E}) = \bigcap_{a \in S} \mathcal{R}^{*E}a \tag{8}$$

Now, any formal concept $(X, S) \in \mathcal{L}(\mathcal{O}, \mathcal{A}, \mathcal{R})$ can be approximated by $\mathcal{R}_{*E}$ and $\mathcal{R}^{*E}$. The *lower* and *upper E-approximation* of (X, S) are defined as

$$(X, S)_{*E} = (\rho(S_{*E}), \lambda\rho(S_{*E})) \in \mathcal{L}(\mathcal{O}, \mathcal{A}, \mathcal{R}_{*E}) \tag{9}$$

$$(X, S)^{*E} = (\rho(S^{*E}), \lambda\rho(S^{*E})) \in \mathcal{L}(\mathcal{O}, \mathcal{A}, \mathcal{R}^{*E}) \tag{10}$$

A *rough concept* of a formal concept $(\mathcal{O}, \mathcal{A}, \mathcal{R})$ in $(\mathcal{O}, E)$ is the collection of concepts which have the same lower and upper E-approximations (roughly equal concepts). Note that approximate contexts of $(\mathcal{O}, \mathcal{A}, \mathcal{R})$ in $(\mathcal{O}, E)$ vary according to the equivalence relation E. In [3] we introduce algorithm A-OSHAM for learning approximate concepts in the framework of rough concept analysis. Essentially, A-OSHAM induces a concept hierarchy in which each induced concept is associated with a pair of its lower and upper approximations. A-OSHAM generates concepts with their approximations recursively and gradually, once a level of the hierarchy is formed the procedure is repeated for each class.

4 Document Clustering based on a Tolerance Rough Set Model

Given a set $\mathcal{D}$ of M full text documents. Our method of generating a hierarchical structure of this document collection consists of two phases. The first

Table 2: Scheme of A-OSHAM approximate conceptual clustering

Input concept hierarchy H and an existing splittable concept C_k.
Result H formed gradually.
Top-level call A-OSHAM(root concept, $\emptyset$).
Variables α is a given threshold.

1. Suppose that $C_{k_1}, ..., C_{k_n}$ are subconcepts of $C_k = (X_k, S_k)$ found so far. While C_k is still splittable, find a new subconcept $C_{k_{n+1}} = (X_{k_{n+1}}, S_{k_{n+1}})$ of C_k and its approximations by doing:
   (a) Find attribute a^* so that $\bigcup_{i=1}^{n} X_{k_i} \cup \rho(\{a^*\})$ is the largest cover of X_k.
   (b) Find the largest attribute set S containing a^* satisfying $\lambda\rho(S) = S$.
   (c) Form subconcept $C_{k_{n+1}}$ with $\rho(S_{k_{n+1}}) = S$ and $X_{k_{n+1}} = \rho(S)$.
   (d) Find a lower approximation and an upper approximation of $C_{k_{n+1}}$ with respect to a chosen equivalence relation E.

   From intersecting subconcepts corresponding to intersections of $\rho(S_{k_{n+1}})$ with extents of existing concepts on H excluding its superconcepts, and find their approximations.
2. Let $X_k^r = X_k \setminus \bigcup_{i=1}^{n+1} X_{k_i}$. If one of the following conditions holds then C_k is considered unsplittable:
   (a) There exist not any attribute set $S \subseteq S_k$ satisfying $\lambda\rho(S) = S$ in X_k.
   (b) $card(X_k^r) \leq \alpha$.
3. Apply A-OSHAM(C_{k_i}, H) to each C_{k_i} formed in the step 1.

---

phase extracts and maps each document into a set of terms, then enriches documents with their approximations by the proposed tolerance rough set model. The second phase groups documents by an agglomerative clustering method using the document approximations.

In the first phase each document d_j is mapped into a list of terms t_i each is assigned a weight that reflects its importance in the document. Denote by $f_{d_j}(t_i)$ the number of occurrences of term t_i in d_j (term frequency), and by $f_{\mathcal{D}}(t_i)$ the number of documents in $\mathcal{D}$ that term t_i occurs in (document frequency). The weights w_{ij} of terms t_i in documents d_j are first calculated by

$$w_{ij} = \begin{cases} (1 + \log(f_{d_j}(t_i))) \times \log \frac{M}{f_{\mathcal{D}}(t_i)} & \text{if } t_i \in d_j, \\ 0 & \text{if } t_i \notin d_j \end{cases} \tag{11}$$

then normalized by vector length as $w_{ij} \leftarrow w_{ij}/\sqrt{\sum_{t_{hj} \in d_j}(w_{hj})^2}$. Each document d_j is represented by its r highest-weighted terms. A usual way is to fix a default value r common for all documents. We denote the document set by $\mathcal{D} = \{d_1, d_2, \ldots, d_M\}$ where $d_j = (t_{1j}, w_{1j}; t_{2j}, w_{2j}; \ldots; t_{rj}, w_{rj})$ and $w_{ij} \in [0, 1]$. The set of all terms from $\mathcal{D}$ is denoted by $\mathcal{T} = \{t_1, t_2, \ldots, t_N\}$. In information retrieval, a query is given the form $Q = (q_1, w_{1q}; q_2, w_{2q}; \ldots; q_s, w_{sq})$ where $q_i \in \mathcal{T}$ and $w_{iq} \in [0, 1]$.

The *tolerance rough set model* (TRSM) aims to enrich the document representation in terms of semantics relatedness by creating tolerance classes of terms in $\mathcal{T}$ and approximations of subsets of documents. The model has the root from rough set models and its extensions [10]. The key idea is among three properties of an equivalence relation R in an universe U used in the original rough set model (reflexive: xRx; symmetric: $xRy \rightarrow yRx$; transitive: $xRy \wedge yRz \rightarrow xRz$ for $\forall x, y, z \in U$), the transitive property does not always hold in natural language processing, information retrieval, and consequently text data mining. In fact, words are better viewed as overlapping classes which can be generated by *tolerance relations* (requiring only reflexive and symmetric properties).

The key issue in formulating a TRSM to represent documents is the identification of tolerance classes of index terms. There are several ways to identify conceptually similar index terms, e.g., human experts, thesaurus, term co-occurrence, etc. We employ the co-occurrence of index terms in all documents from $\mathcal{D}$ to determine a tolerance relation and tolerance classes. The co-occurrence of index terms is chosen for the following reasons: (i) it gives a meaningful interpretation in the context of information retrieval about the dependency and the semantic relation of index terms, and (ii) it is relatively simple and computationally efficient. Note that the co-occurrence of index terms is not transitive and cannot be used automatically to identify equivalence classes. Denote by $f_{\mathcal{D}}(t_i, t_j)$ the number of documents in $\mathcal{D}$ in which two index terms t_i and t_j co-occur. We define an uncertainty function I depending on a threshold θ as

$$I_\theta(t_i) = \{t_j \mid f_{\mathcal{D}}(t_i, t_j) \geq \theta\} \cup \{t_i\} \tag{12}$$

It is clear that the function I_θ defined above satisfies the condition of $t_i \in I_\theta(t_i)$ and $t_j \in I_\theta(t_i)$ iff $t_i \in I_\theta(t_j)$ for any $t_i, t_j \in \mathcal{T}$, and so I_θ is both

Table 3: The TRSM nonhierarchical clustering algorithm

Input	The set $\mathcal{D}$ of documents and the number K of clusters.
Result	K clusters of $\mathcal{D}$ associated with cluster membership of each document.

1. Determine the initial representatives $R_1, R_2, ..., R_K$ of clusters $C_1, C_2, ..., C_K$ as K randomly selected documents in $\mathcal{D}$.
2. For each $d_j \in \mathcal{D}$, calculate the similarity $S(\mathcal{U}(\mathcal{R}, d_j), R_k)$ between its upper approximation $\mathcal{U}(\mathcal{R}, d_j)$ and the cluster representative $R_k, k = 1, ..., K$. If this similarity is greater than a given threshold, assign d_j to C_k and take this similarity value as the cluster membership $m(d_j)$ of d_j in C_k.
3. For each cluster C_k, re-determine its representative R_k.
4. Repeat steps 2 and 3 until there is little or no change in cluster membership during a pass through $\mathcal{D}$.
5. Denote by d_u an unclassified document after steps 2, 3, 4 and by NN(d_u) its nearest neighbor document (with non-zero similarity) in formed clusters. Assign d_u into the cluster that contains NN(d_u), and determine the cluster membership of d_u in this cluster as the product $m(d_u) = m(\text{NN}(d_u)) \times S(\mathcal{U}(\mathcal{R}, d_u), \mathcal{U}(\mathcal{R}, \text{NN}(d_u)))$. Re-determine the representatives R_k, for $k = 1, ..., K$.

reflexive and symmetric. This function corresponds to a tolerance relation $\mathcal{I} \subseteq \mathcal{T} \times \mathcal{T}$ that $t_i \mathcal{I} t_j$ iff $t_j \in I_\theta(t_i)$, and $I_\theta(t_i)$ is the tolerance class of index term t_i.

A vague inclusion function ν, which determines how much X is included in Y, is defined as

$$\nu(X, Y) = \frac{|X \cap Y|}{|X|} \tag{13}$$

This function is clearly monotonous with respect to the second argument. Using this function the membership function, introduced by Pawlak [10], a similar notion as that in fuzzy sets, μ for $t_i \in \mathcal{T}, X \subseteq \mathcal{T}$ can be defined as

$$\mu(t_i, X) = \nu(I_\theta(t_i), X) = \frac{|I_\theta(t_i) \cap X|}{|I_\theta(t_i)|} \tag{14}$$

With these definitions we can define a tolerance space as $\mathcal{R} = (\mathcal{T}, I, \nu, P)$ in which the *lower approximation* $\mathcal{L}(\mathcal{R}, X)$ and the *upper approximation*

Table 4: TRSM-based hierarchical agglomerative clustering algorithm

---

Input A collection of M documents $\mathcal{D} = \{d_1, d_2, \ldots, d_M\}$
Result Hierarchical structure of $\mathcal{D}$

Given: a collection of M documents $\mathcal{D} = \{d_1, d_2, \ldots, d_M\}$
a similarity measure $sim : \mathcal{P}(\mathcal{D}) \times \mathcal{P}(\mathcal{D}) \rightarrow R$
for $j = 1$ **to** M **do**
$C_j = \{d_j\}$ **end**
$H = \{C_1, C_2, , \ldots, C_M\}$
$i = M + 1$
while $|H| > 1$
$(C_{n_1}, C_{n_2}) = \text{argmax}_{(C_u, C_v) \in H \times H} sim(\mathcal{U}(\mathcal{R}, C_u), \mathcal{U}(\mathcal{R}, C_v)$
$C_i = C_{n_1} \cup C_{n_2}$
$H = (H \setminus \{C_{n_1}, C_{n_2}\}) \cup \{C_i\}$
$i = i + 1$

$\mathcal{U}(\mathcal{R}, X)$ in $\mathcal{R}$ of any subset $X \subseteq \mathcal{T}$ can be defined as

$$\mathcal{L}(\mathcal{R}, X) = \{t_i \in \mathcal{T} \mid \nu(I_\theta(t_i), X) = 1\} \tag{15}$$

$$\mathcal{U}(\mathcal{R}, X) = \{t_i \in \mathcal{T} \mid \nu(I_\theta(t_i), X) > 0\} \tag{16}$$

The term-weighting method defined by Eq. (11) is extended to define weights for terms in the upper approximation $\mathcal{U}(\mathcal{R}, d_j)$ of d_j. It ensures that each term in the upper approximation of d_j but not in d_j has a weight smaller than the weight of any term in d_j.

$$w_{ij} = \begin{cases} (1 + \log(f_{d_j}(t_i))) \times \log \frac{M}{f_{\mathcal{D}}(t_i)} & \text{if } t_i \in d_j, \\ \min_{t_h \in d_j} w_{hj} \times \frac{\log(M/f_{\mathcal{D}}(t_i))}{1+\log(M/f_{\mathcal{D}}(t_i))} & \text{if } t_i \in \mathcal{U}(\mathcal{R}, d_j) \setminus d_j \\ 0 & \text{if } t_i \notin \mathcal{U}(\mathcal{R}, d_j) \end{cases} \tag{17}$$

The vector length normalization is then applied to the upper approximation $\mathcal{U}(\mathcal{R}, d_j)$ of d_j. Note that the normalization is done when considering a given set of index terms.

Figure 3 and Figure 4 describe two general TRSM-based nonhierarchical and hierarchical clustering algorithm. The TRSM-based nonhierarchical clustering algorithm can be considered as a reallocation clustering method to form K clusters of a collection $\mathcal{D}$ of M documents. The main point of the TRSM-based hierarchical clustering algorithm is at each merging step it uses upper approximations of documents in finding two closest clusters to merge. Several variants of agglomerative clustering can be applied, such single-link or complete-link clustering. As documents are represented as length-normalized vectors and when cosine similarity measure is used, an efficient alternative is to employ the group-average agglomerative clustering. The group-average clustering can avoid the elongated and straggling clusters produced by single-link clustering, and can avoid the high cost of complete link clustering. In fact, it allows using cluster representatives to calculate the similarity between two clusters instead of averaging similarities of all document pairs each belong to one cluster [8]. In such a case, the complexity of computing average similarity would be $O(N^2)$. Careful evaluation and validation of clustering quality are given in [5] and [6]. The results show that tolerance rough set model and TRSM-based clustering algorithms can be used to improve the effectiveness and efficiency in information retrieval and text analysis.

References

[1] Ganter, B., Wille, R., *Formal Concept Analysis: Mathematical Foundations*, Springer-Verlag, 1999.

[2] Ho, T.B., Discovering and Using Knowledge From Unsupervised Data. *Decision Support Systems*, Elsevier Science, Vol. 21, No. 1, 1997, 27–41.

[3] Ho, T.B., "Two approaches to the representation and interpretation of concepts in concept lattices", *Information Modelling and Knowledge Bases XI*, IOS Press, 2000, 12-25.

[4] Ho, T. B. and Funakoshi, K., "Information retrieval using rough sets", *Journal of Japanese Society for Artificial Intelligence*, Vol. 13, No. 3, 1998, 424–433.

[5] Ho, T. B. and Nguyen, N.B., "Nonhierarchical Document Clustering by Tolerance Rough Set Model", *International Journal of Intelligent Systems*, Vol. 17 (2002), No. 2, 199–212.

[6] Kawasaki, S., Nguyen, N.B., and Ho, T. B., "Hierarchical Document Clustering Based on Tolerance Rough Set Model", *Fourth European Conference on Principles of Data Mining and Knowledge Discovery*, September 2000, Lyon. Lecture Notes in Artificial Intelligence 1910, Springer, 458–463

[7] Kent, R.E., "Rough concept analysis", *Rough Sets, Fuzzy Sets and Knowledge Discovery*, Springer-Verlag, 1994, 248–255.

[8] Manning, C. D. and Schutze, H., *Foundations of Statistical Natural Language Processing*, The MIT Press, 1999.

[9] Michalski, R.S. and Stepp, R.E., "Learning from observation: Conceptual learning", *Machine Learning: An Artificial Intelligence Approach*, Vol. 1, R. S. Michalski, J. G. Carbonelle, T. M. Michell (Eds.), Morgan Kaufmann, 1983, 331–363.

[10] Pawlak, Z., *Rough Sets: Theoretical Aspects of Reasoning about Data*, Kluwer Academic Publishers, 1991.

[11] Wille, R., "Restructuring lattice theory: An approach based on hierarchies of concepts", *Ordered Sets*, I. Rival (Ed.), Reidel, 1982, 445–470.

A Simple Heuristic Method for the Min-Cut k-Balanced Partitioning Problem

Lelia Blin and Ivan Lavallee

Abstract *We consider the problem of k-partitioning a graph into k balanced subsets that minimizes the number of crossing weighted edges Previously, Maximum flow approach, iterative improvement, geometric representation were given to solve this problem. In this paper we propose a balanced k-partition heuristic based on a max spanning tree method*

Keywords: Bipartition, k-Partition, Min-Cut, Balanced, Merging.

1 Introduction

Graph connectivity is one of the classical subjects in graph theoiy The paititioning of the grapbs is an important problem that has extensive applications in mmy areas, including scientific computing, VLSI design [AK95], data mining [MJHS96], geo-glaphical information systems, operation research, and task scheduling. In circuit design, working with VLSI-CAD systems yields a high level of complexity because the number of electronic circuits increases very fast due to the technology improvements. Current designs are in the order of 100.000 cells and it is expected to reach 1 000.000 cells in next years. To give an idea of the involved complexity when working in design, actually 300.000 cells requires 8 hours of processing time running in parallel under 128 networked Sparc stations. The system must deal with the combinatorial explosion of al1 the possible solutions and it is very important to find new efficient heuristics.

Much heuristics are known as interchange methods and are based on iteratively improving a series of partitions. In general, the quality of the final solution depends on the quality of the initial partition. Moreover, iterative exchange approaches cam easily can trapped in local minima. In contrast, we propose a simple and efficient heuristic method usine a global optimization criterion that promises a better performance.

The rest of the paper is organized as follows : section two introduces the problem, on section three we discuss the state-of-the-art, section four describes our heuristic approach and illustrates with an example.

2 The problem

- **Instance:** Given $G = (V, E)$ undirected graph [Ber73], weights $w(e) \in Z^+$ for each $e \in E$. Let V the vertex set, $|V| = n$, and the edges set E, $|E| = m$. A cuted edge is an edge (u, v) where $u \in V_i$, and $v6 \notin V_i$. The cut set $CUT(E)$ is the set with all the cuted edges in G. k is given as a positive integer.

- **Question:** Is there a k-partition of V into disjoint sets $V_1, V_2, ..., V_k$ such that minimize
$$\sum_{E \in CUT(E)} w(e)$$

This problem is known to be NP-hard[GJ79], thus, many heuristics have been developed to obtain suitable partitions.

3 The state of the art

The k-partitioning problem is most frequently solved by recursive bipartition. That is, we first obtain a 2-partitioning (bipartitioning) of V, and then we recursively obtain a 2-partitioning of each resulting partition. After many phases, graph G is partitioned into k partitions. Thus, the problem of performing a k partitioning is reduced to that of performing a sequence of bipartitions. Consequently, in the literature, many heuristics have been proposed in the context of Bipartitioning. Whereas, one of the original issues in this work is to directly construct k-partitions without making use of bi-partitioning.

There are two types of approaches based on a Min-Cut Graph Bipartitioning: The first one only seeks one minimal cut in the bipartition and the second one also seeks a minimal eut cut add a constraint: the balance constraint, the disjoints subsets of V must have the same size Min-Cut.

3.1 Bi-Partitioning: MC2P

The usual approach to solve this problem is to use its close relationship with the maximum-flow problem[FF62]. In this field there are very simple heulistic methods proposed [SW97]. However, it was overlooked as a viable approach for circuit partitioning due to the following reasons:

- The two components obtained by the network max-flow min-cut technique are not necesarly balanced.
- Although, a balanced eut can be achieved by repeatedly applying min-cut to the larger component, this method can possibly incur n max-flow computations.
- The traditional network-flow technique works on graphs, but hypergraphs are more accurate models for circuit netlists than graphs.

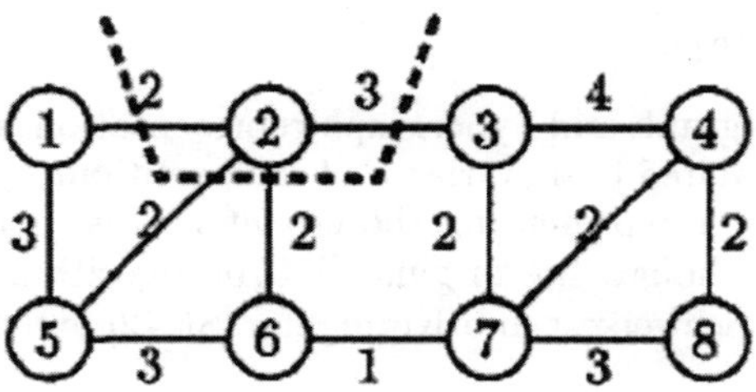

Figure 1: Example of Stoer et Wagner heuristic methods [SW97]. The min cut is 9, partition is {2}{1,3,4,5,7,8}

3.2 Min-Cut Balanced Bi-Partioning: MCB2P

The standard bipartitioning approach is iterative improvement based on the Kernighan-Lin algorithm [KL70], which was later improved by Fiduccia-Mattheyses [FM82].

In 1970, Kerighan and Lin [KL70] introduced what is often described as the first graph bisection heuristic. Their algorithm begins whith some initial solution (A^*, B^*). The KL method uses a par-swap neighborhood structure and proceeds in a series of passes.

Suppose that (A^*, B^*) is a minimum cost bi-partition and let (A, B) be any arbitrary bi-partition. Then there are two subsets with $X \in A$ and $Y \in B$ where $|X| = |y| \leq n/2$. Thus, interchanging X and Y may produce A^* and B^* as shown below X and Y are found by interchanging two nodes in each iteration.

Example of Khernighan and Lin heuristic method.

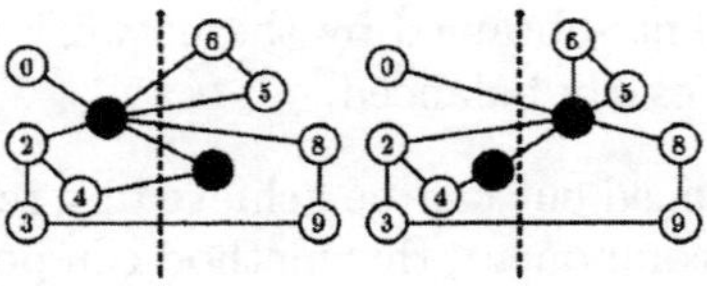

Figure 2: While interchanging nodes 1 and 7, the cut passes from 6 cuted edges to 4)

3.3 Hypergraph

There are various graph and hypergraph representations of the circuit netlist (VLSI), and formulated basic variants of the partitioning problem. The most common method for representing the circuit netlsist connections (VLSI) is a hypergraph. Although the original [KL70] algorithm only to undirected weighted graphs, Schweikert and Kernighan [SK72] extend [KL70] to hypergraphs.

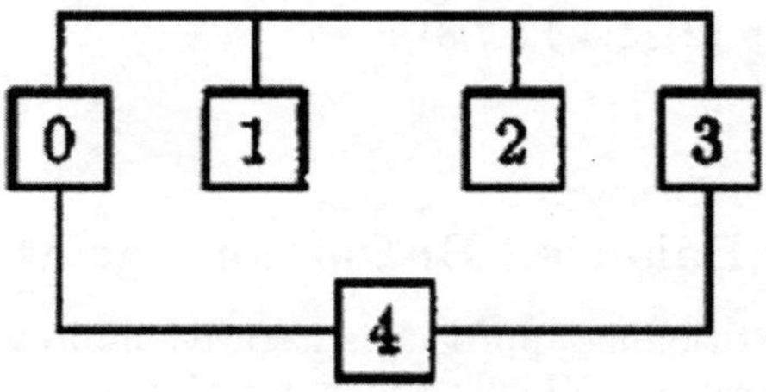

Figure 3: Circuit

However, iterative improvement methods was overlooked as a viable approach for circuit partitioning due to the following reasons:

- the quality of the final solution depends on the quality of the initial partition.
- the iterative method is very expensive and far from their practical use (a good implementation KL yields a complexity in the oider of $0(n^2)$.
- It is not possible to make k-partitioning directly.

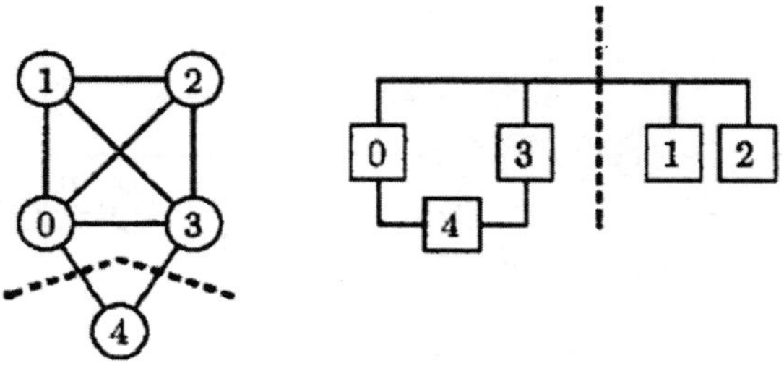

Figure 4: Graph representation of the circuit, circuit representation by hypergraph

3.4 Geometric Representation

In this part, we discusses methods that construct a gometric representation of the partitioning problem via constructions as such a 1-dimensional linear ordering or multi-dimensional vector space. Such a represation offers possibilities for geometric approaches to solve problems that are intractable for general graphs. Spectral methods are commonly used to construct geometric representations, due to their ability to capture global netlist information.

Illustration of Hall's Quadratic Placement [Hal70] The significance of hall's result is that it provides the optimal non-disciete solution for Min-Cut Bipartitioning. Given an nxn symmetric connection matrix $C = (c_{ij})$. Eigenvaiues are computed by starting from this matrix. The eigenvectors associated E_1, E_2, E_3, E_4 with the eigenvalues are obtained $\lambda = (0.0, 0.586, 2.0,$ $3.414)$. To find the cut there are two possible choices:

First choice Placement in one dimension. By using the two smallet eigenvalues: i.e. the vectors E_1 and E_2 where the nodes are placed on the line according to the coordinates of E_2 (E_1 line). The cut is at the center of the line and it is the optimum because only one edge is cut.

Second choice Placement in 2 dimensions $E_1 = 0$ is a line, it cannot thus be used. That is why the two others smaller values of λ are used, i.e the eigenvectors E_2 and E_3 where the nodes are placed following X coordinated and Y coordinated. The cut passes along the x-axis and this is the optimum because only one edge is cut.

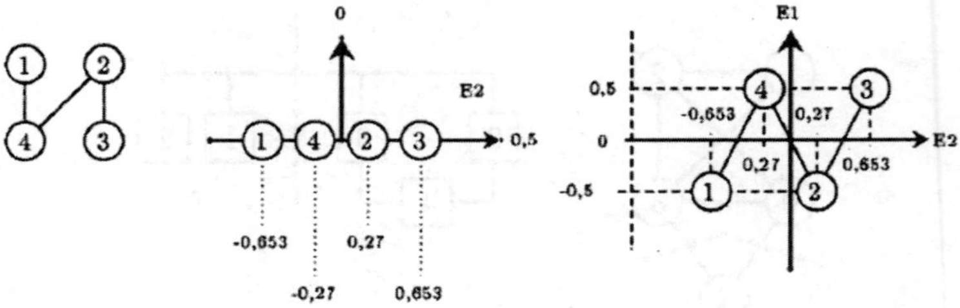

Figure 5: Method of placement: The graph, one dimension, and two dimensions

3.5 Replication

Replication modules can reduce the cutsize, and are paiticulary useful for FPGA paititioning since many device architectures seem more I/O-limited or interconnected-limited than logic-limited. Replication can also reduce the number of interchip wires along a given path, increasing system performance. The Min-Cut Replication Problem of Hwang and El Gamal [HG92] seeks a collection of subsets of modules $C_{ij}^*|C_{ij}^*, 1 \leq i, j \leq k$ that minimizes $F(P^{k*})$, where P^{k*} is the partitioning that results when each subsets C_{ij}^* is replicated from C_i to C_j. Hwang and El Gamal implicitly assume each C_i contains a subset I_i of primary imputs that cannot be replicated. Consider the directed graph shown in figure 6(a), in which module v represents an N-input decoder circuit. The cut shown has size 2^N but if the decoder v is replicated as in 6(h), every one of these 2^N edges will become uncut (however, N new edges will be cut). The following rules are used to modify the edge set E when $v \in C_h$ is replicated into $v' \in C_l$. In [HG95], Hwang and El Gamal showed how to modify their flow network to solve min-cut replication for hypergraphs with signal information

We have presented a non-exhaustive review of the existing methods in this field. These methods have several shelfs. The first shelf is that they are not easy to understand and to implement. The second shelf it is that they are not very powerful, the execution time becomes prohibitory for the circuits'size interesting. The last major problem, is that they do not propose k-partition but of the reiteration of bipartition, but in the reality of the VLSI, it is necessary to tackle k-partition. These heuristics have execution times which make them of not use in state, their repetition to obtain k-partition only makes worst this observation.

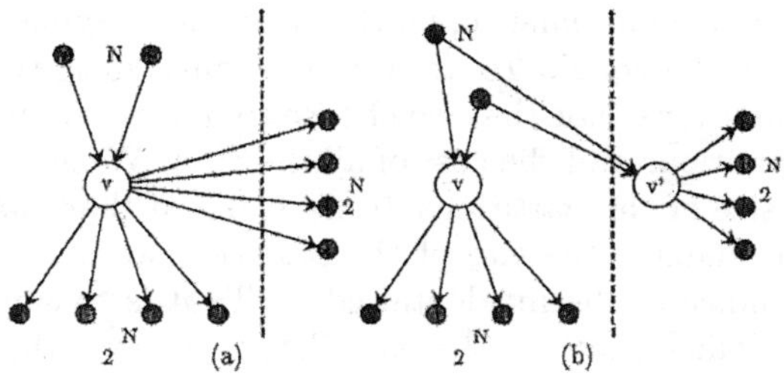

Figure 6: Replication

4 Our heuristic

We locate our heuristic in the min-cut balanced k-partition (MCBkP). The main idea of our heuristic is simple. The edges of maximum weight should not be found in the cut-set CUT(E), that minimize the cut-set. Moreover the weights of the edges will be small in the unit of cut, the more their will tend towards a minimum.

The search of maximum edges can be made by the search of the maximum spanning tree. We use the Sollin algorithm [Sol63] for finding the maximum spanning tree. This algorithm does not require preprocessing of the edges, save computing time, and adopts a "bottom-up" strategy. Moreover we build the tree from several forests, which authorizes a parallelism [Lav91] and this principle is the base for k-partitioning. The general principle is the Borukva phase [Bur26]. By merging node sets, we obtain a new graph and we iterate the strategy on the new graph. During the heuritic method, a sequence of successively smaller graphs is constructed.

Our heuristic given $G = (V, E)$ undirected graph [Ber73], weights $w(e) in Z^+$ for each $e \in E$. Let V the vertex set, $|V| = n$, and the edges set E, $|E| = m$. A cuted edge is an edge (u, v) where $u \in V_i$, and $v \notin V_i$. The cut set $CUT(E)$ is the set with all the cuted edges in G. k is given as a positive integer, it is the number of partition. We associate to each node a state; a node is marked if already visited, unmarked if not. At the beginning each node is unmarked.

It is easy to understand the heuristic. The unmarked vertices are visited in a random order, and for each vertex v, the edge incident on v with the highest edge-weight is selected, $e = (Max\ w(\Gamma[v]) = (v, u)$, where $\Gamma[v]$ stands for the edges incidents of v. Once all vertices have been marked, the unselected edges are moved, and each one of the connected components of the resulting graph becomes a set of vertices to be merged together.

Although not being completely unbalanced, the partitions obtained do not satisfy the criterion of balance, ie, the constraint saying that the sizes of the partitions must be equal. To answer this constraint, we should manage the size of the partitions. For that we choose to make a compromise between the size of the partitions and the cost of all the cuts. We accept an unbalance r (fixed) on the size of the partitions, $0 < r < 0.5$, if that enables us to have a better cut. To manage the size of the partition we insert a test into the moment of the choice of the incidental edge. That is to say V_i a unmarked node for an unspecified graph, $|v_i| = p_i$. That is to say e the incidental edge with V_i stsonger weight, $e_1 = (Max\ w(\Gamma[V_i])) = (V_i, v_{j1})$ and $|v_{j1}| = p_{j1}$. If $p_i + p_j < (\frac{|V|}{k} + r\frac{|V|}{k})$ one selects e, if not we select the following node with the strongest weight, $e_2 = (Max\ w(\Gamma[V_i]) \neq w(e_1)) = (V_i, V_{j2})$, and so on...

4.1 Example

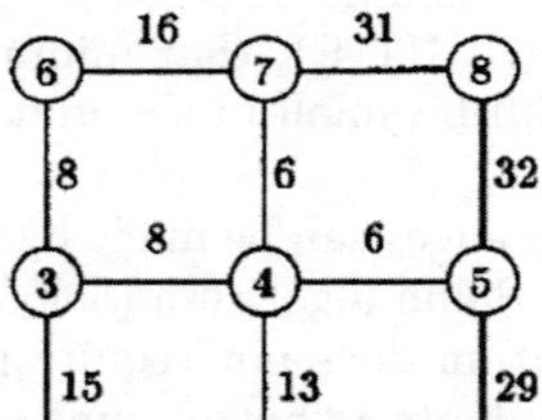

Figure 7: Initial graph

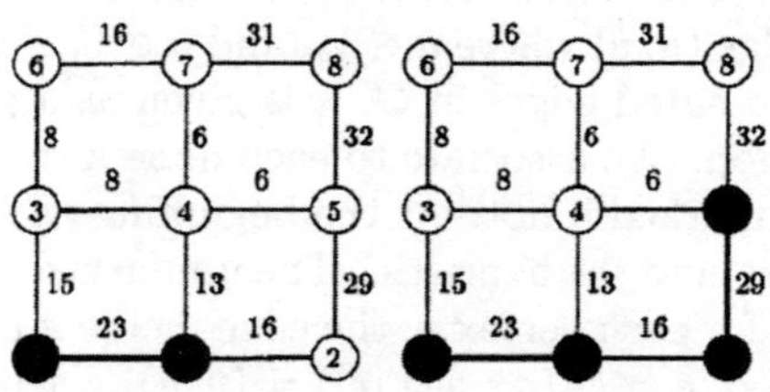

Figure 8: First phase:first and second step

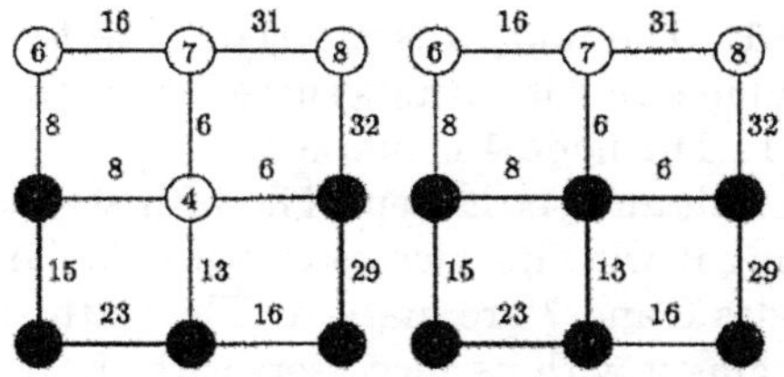

Figure 9: First phase:third and fourth step

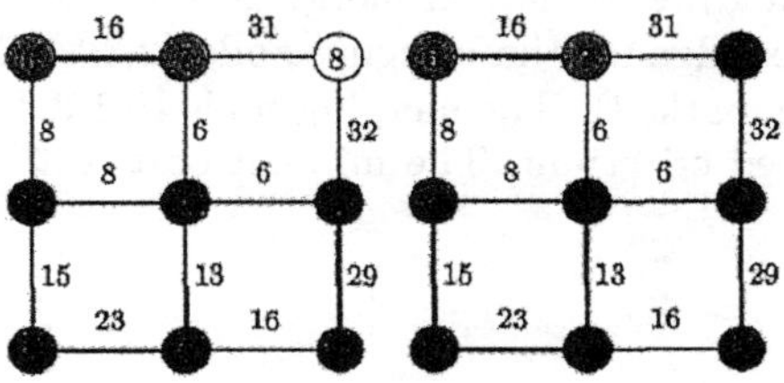

Figure 10: First phase:fifth and sixth step

As on Figure 7, in a graph $G = (V, E)$, we seek a balanced bi-partition. The Figures 8, 9, 10 illustre the first Phase. As on Figure 8 first and second step. The first step choose the node unmarked 0 and merging it with its successors with the largest weighted edge, it is the node 1 Nodes 0 and 1 are marked. The second step choose the node unmarked 2 and merging it with its successors whith the largest weighted edge, it is the node 5 Nodes 2 and 5 are marked.

As on Figure 9 third and fourth step. The third step choose the node unmarked 3 and merging it with its successors with the largest weighted

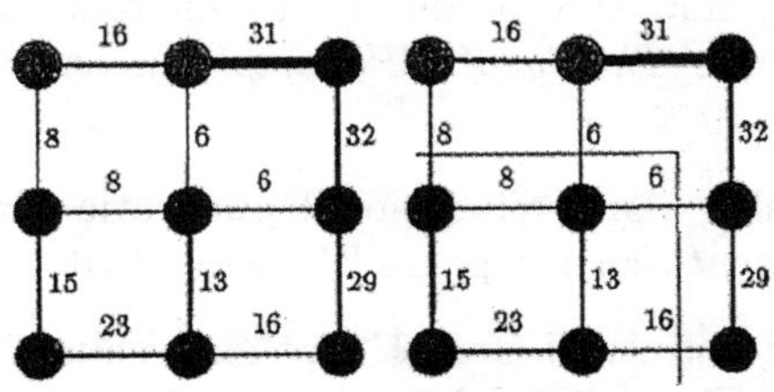

edge, it is the node 1. The node 3 is marked. The fourth step choose the node unmarked 4 and merging it with its successors with the largest weighted edge, it is the node 1. The node 4 is marked.

As on Figure 10 fifth and sixth step. The fifth step choose the node unmarked 6 and merging it with its successors with the largest weighted edge, it is the node 7. Nodes 6 and 7 are marked. The sixth step choose the node unmarked 8 and merging it with its successors with the largest weighted edge, it is the node 5. The node 8 is marked. The figure 4.1 illustres the second phase and the min-cut balanced bipartition. For start the second phase, we unmarked the merging nodes. The second phase have one step, which corresponds to the merging the unmarked-merging-nodes (2,5,8) and the merging nodes (6,7), the max edge for the merging-node (2,5,8) (Now merging-nodes (2,5,8) and (6,7) are marked). The merging-node (0,1,3,4) is not use, because it satisfy the balanced criterium. The min-cut cost is 33.

References

[AK95] Charles J. Alpert and Andrew B. Kahng. Recent directions in netlist partitioning. *Integration, the VLSI Journal*, 19, 1995.

[Ber73] *Graphes et hypergraphes.* North-Holland, Amsterdam, 1973.

[Bur26] O. Burukva. O jistm problmu minimàln im". *Pràce Mor. Prirodoved Spol. v Brne Acta Societ. Scient. Natur Moravicae)*, pages 37–58, 1926.

[FF62] L.R Ford and D.R Fulkerson. Flows in networks. *New Jersey: Princeton University Press*, page 11, 1962.

[FM82] C.M. Fiduccia and R. M. Mattheyses. A linear-time heuristic for improving network partitions. *Proc. 19th ACM/IEEE Design Automation Conf.*, pages 175–181, 1982.

[GJ79] M. Garey and D.S. Johnson. *Computers and Intractability: A guide to the Theory of NP-Completeness.* San Fransisco,W.H. Freeman, 1979.

[Hal70] K.M. Hall. An r-dimensional quadratic placement algorithm. *Management Science*, pages 219–229, 1970.

[HG92] J. Hwang and A. El Gamal. Optimal replication for min-cut partitioning. *Proc. IEEE Int. Conf. Computer-Aided Design*, pages 432–435, Nov 1992.

[HG95] J. Hwang and A. El Gamal. Min-cut replication in partitioned network. *IEEE Trans. Computer-Aided Design*, 14:96–106, Jan 1995.

[KL70] B.W Kernighan and S. Lin. En efficient heuristic procedure for partitioning graph. *Belle Systems Technical Journal*, pages 291–307, 1970.

[Lav91] *Algorithmique parallle et distribue.* Hermes, 1991.

[MJHS96] B. Mobasher, N. Jain, E.H. Han, and J. Srivastava. Web mining: Pattern discovery from world wide web translation. Technical report, Departement of Computer Science, University of Minnesota, Minneapolis, 1996.

[SK72] D.G. Schweikert and B.W Kernighan. A proper model for the partitioning of electonical circuits. *Bell Telephone Laboratories Incorporated, New Jersey*, 1972.

[Sol63] M. Sollin. Expos du sminaire de c. berge, ihp, 1961, repris in extenso dans. *Mthodes et modles de la recherche Oprationnelle, Dunod*, pages 33–45, 1963.

[SW97] M. Stoer and F. Wagner. A simple min-cut algorithm. *Journal of the ACM*, 44(4):585–591, Jul 1997.

LONGEST CYCLES AND RESTGRAPH IN MAXIMAL NONHAMILTONIAN GRAPHS

Vu Dinh Hoa

Faculty of Information Technology
Pedagogical University of Hanoi
136 duong Xuan Thuy Ha noi Vietnam
e-mail: hoavd@fpt.com.vn

Abstract. A graph is called a maximal nonhamiltonian graph if it is nonhamiltonian and will be hamiltonian by adding any new edge. The following conjecture was presented by Erdös that if C is a longest cycle in a maximalnonhamiltonian graph G then $G-C$ is a complete graph. In this paper we prove that the conjecture is true for some classes of graphs.

1 Terminology and Notation

We consider only undirected graphs without loop or multiple edges. Our terminology is standard except as indicated. We begin by introducing some definitions and notation. Let $\omega(G)$ denote the number of components of a graph G. Following Chvátal [2] a graph G is *1-tough* if $|S| \geq \omega(G - S)$ for any nonempty subset S of the vertex set $V(G)$ of G. G is called *nontough* if there exists a nonempty subset S such that $|S| < \omega(G - S)$. We shall denote the number of vertices of G by n. A graph G is called a *hamiltonian* graph if G contains a hamiltonian cycle, i. e. a cycle of length n. Otherwise, G is *nonhamiltonian.* Similarly, a path is called a hamiltonian path if it contains all vertices in G. A graph G is called a *maximalnonhamiltonian* graph if G is nonhamiltonian and will be hamiltonian by adding any new edge. Clearly, for every nonadjacent vertices u and v in a maximalnonhamiltonian graph G there is a hamiltonian path joining u with v. A cycle C of G is a *dominating* cycle if $G - C$ is an edgeless graph and G is called *dominated by* C or *dominable.* The length $\ell(C)$ of a longest cycle C in a graph G, called the *circumference* of G, will be denoted by $c(G)$. For $k \leq \alpha$ we denote by σ_k the minimum value of the degree sum of any k pairwise nonadjacent vertices. For $k > \alpha$ we set $\sigma_k = \infty$. Instead of σ_1 we use the more common notation δ.

2 Results and Conjecture

We begin with a well-known theorem of Nash-Williams [4].

Theorem 1 *Let G be a 2-connected graph on n vertices with $\delta \geq (n+2)/3$. Then every longest cycle in G is a dominating cycle.*

Bigalke and Jung [1] improved Theorem 1 under the assumption that G is 1-tough.

Theorem 2 *Let G be a 1-tough graph on n vertices with $\delta \geq n/3$. Then every longest cycle in G is a dominating cycle.*

The class of graphs with $\delta \geq \frac{n}{3}$ was also studied by Enomoto et other in [3]. Let C be a longest cycle in a nonhamiltonian graph G. In the proof technics for the existence of a hamiltonian cycle we often ask for the rest graph $G - C$. Clearly, it is very usefull if we know about $G - C$ for any longest cycle C in a maximalnonhamiltonian graph G. For some special cases we can describe the rest graph $G - C$. A such result is proved in [6].

Theorem 3 *Let G be a 2-connected graph on n vertices with $\delta \geq n/3$ and C be a longest cycle in it. Then $G - C$ is either a complete graph or an edgeless graph.*

It was also shown in [6] that if $\delta \geq n/3$ and G is 2-connected then $G - C$ is only dependent on G, namely:

Theorem 4 *Let G be a 2-connected graph on $n \geq 3$ vertices with $\delta \geq n/3$. Then the following properties are equivalent:*
(a) G is a dominable graph,
(b) every longest cycle in G is a dominating cycle,
(c) $G \notin \mathfrak{R}$ for a class $\mathfrak{R}$ of special graphs.

The definition of the class $\mathfrak{R}$ shows that Theorems 4 is a generalization of Theorem 1 and Theorem 2. The class $\mathfrak{R}$ is the union of the classes of graphs $\mathfrak{R}_1, \ldots, \mathfrak{R}_5$ and it takes only a polynomial time to know whether a given graph $G \in \mathfrak{R}$ or not.
$\mathfrak{R}_1$ is the class of graphs on $3r$ vertices and with minimal degree $\delta = r \geq 3$, which are isomorphic to a graph G, where

$$2K_{r-1} + \bar{K}_2 \subseteq G \subseteq (2K_{r-1} \cup K_r) + K_2;$$

$\mathfrak{R}_2$ the class of graphs on 15 vertices and with minimal degree $\delta = 5$, which are isomorphic to a graph G, where

$$4K_3 + \bar{K}_3 \subseteq G \subseteq 4K_3 + K_3;$$

$\mathfrak{R}_3$ the class of graphs on $3r$ vertices and with minimal degree $\delta = r \geq 3$, which are isomorphic to a graph G, where

$$(r-1)K_2 + \bar{K}_{r-1} \subseteq G \subseteq ((r-1)K_2 \cup K_3) + K_{r-1};$$

$\mathfrak{R}_4$ the class of graphs on $3r+2$ vertices and with minimal degree $\delta = r+1$, which are isomorphic to a graph G, where

$$3K_r + \bar{K}_2 \subseteq G \subseteq 3K_r + K_2;$$

$\Re_5$ the class of graphs on $3r+2$ vertices and with minimal degree $\delta = r+1 \geq 3$, which are isomorphic to a graph G, where

$$(r+1)K_2 + \bar{K}_r \subseteq G \subseteq (r+1)K_2 + K_r.$$

Note that all graphs of $\Re$ are nontough graphs, we can easily see that the graph $G - C$ is a complete graph for any graph $G \in \Re$ and every longest cycle C in G. The following conjecture was posed by Vu Dinh Hoa and presented by Erdös in the *"Second Kraków Conference of Graph Theory"* (September 1994).

Conjeture 1 *If G is a maximalnonhamiltonian graph and C is any longest cycle in G, then $G - C$ is a complete graph.*

In the following, we show that the conjecture is true for nontough maximalnonhamiltonian graphs.

Theorem 5 *Let C be a longest cycle in a nontough maximalnonhamiltonian graph G, then $G - C$ is a complete graph.*

Proof. Since G is nontough, there exists a set S of vertices such that $\omega(G - S) \geq |S| + 1$. Let $s = |S|$ and G_1, G_2 .. G_m the components of $G - S$ with $m \geq s+1$. Clearly, G_1, G_2 ... G_m are complete graphs and $m = s+1$ since G is a maximalnonhamiltonian graph. Otherwise, we can add a new edge in some G_i or a new edge joining two of the components of $G - S$ to obtain a new nonhamiltonian graph with more edges. Similarly, S is a complete graph and every vertex of S is joining with any vertex of $G - S$. Now, we can easily see that for every longest cycle C the graph $G - C$ is one of the components G_1, G_2, ... G_m, i. e. a complete graph.

Now we will show that the same result holds for the class of graphs with $\delta \geq \frac{n}{3}$. We denote by $p(G)$ the length of a longest path in G. In [5], Van den Heuvel proved the following result.

Lemma 1 *(Corollary 6.8 in [5]) Let G be a 1-tough graph on $n \geq 3$ vertices such that $\sigma_3 \geq n$. Then G satisfies $c(G) \geq p(G)$.*

Thus, we conclude the following result.

Theorem 6 *Let G be a maximalnonhamiltonian graph on $n \geq 3$ vertices such that $\delta \geq \frac{n}{3}$ and C be a longest cycle in G, then $G - C$ is a complete graph.*

Proof. If G is nontough graph, then $G-C$ is a complete graph because of theorem 5. Otherwise, G is a 1-tough graph and G satisfies the condition $\sigma_3 \geq n$. By lemma 1, we have $c(G) \geq p(G)$. But, since G is maximalnonhamiltonian graph, $p(G) \geq n-1$. Thus $G-C$ has at most one vertex and therefore $G-C$ is a complete graph.

By the similarly proof, we have the same result for graph G with $\sigma_3 \geq n$.

Theorem 7 *Let G be a maximalnonhamiltonian graph on $n \geq 3$ vertices such that $\sigma_3 \geq n$ and C be a longest cycle in a graph G. Then $G - C$ is a complete graph.*

References

1. A. Bigalke and H. A. Jung, *Über Hamiltonsche Kreise und unabhängige Ecken in Graphen,* Monatsh. Mathematics **88** (1979), 195-210
2. V. Chvátal, *Tough graphs and hamiltonian circuit,* Discrete Math. **5** (1973) 215 - 228.
3. H. Enomoto, A. Kaneko and Zs. Tuza, *P_3-factors and Covering Cycles in Graphs of Minimum Degree n/3*, Colloquia Mathematica Societatis János Bolyai, 52. Combinatorics, Eger (Hungary), 1987.
4. C. St. J. A. Nash - Williams, *Edge- disjoint hamiltonian circuits in graphs with vertices of large valency,* In "Studies in Pure Mathematics", L. Mirsky ed. Academic Press, London 1971, 157-183
5. Van den Heuvel, J., *Degree and Toughness Condition for Cycles in Graphs,* Ph. D. Thesis University of Twente, Netherland 1994.
6. Vu Dinh Hoa, *Ein Struktursatz für 2-fach zusammenhängende Graphen mit großer Minimalvalenz* Math. Nachr. **128** (1986), 151-160.

Deterministic and Nondeterministic Directable Automata

Masami Ito
Department of Mathematics
Kyoto Sangyo University, Kyoto 603-8555, Japan
email: ito@kyoto-su.ac.jp

Abstract

This paper is a survey article to introduce some results on deterministic and nondeterministic directable automata and their related languages.

1 Introduction

Let X be a nonempty finite set, called an *alphabet*. An element of X is called a *letter*. By X^*, we denote the free monoid generated by X. Let $X^+ = X^* \setminus \{\epsilon\}$ where ϵ denotes the empty word of X^*. For the sake of simplicity, if $X = \{a\}$, then we write a^+ and a^* instead of $\{a\}^+$ and $\{a\}^*$, respectively. Let $L \subseteq X^*$. Then L is called a *language* over X. If $L \subseteq X^*$, then L^+ denotes the set of all concatenations of words in L and $L^* = L^+ \cup \{\epsilon\}$. In particular, if $L = \{w\}$, then we write w^+ and w^* instead of $\{w\}^+$ and $\{w\}^*$, respectively. Let $u \in X^*$. Then u is called a *word* over X. If $u \in X^*$, then $|u|$ denotes the length of u, i.e. the number of letters appearing in u. Notice that we also denote the cardinality of a finite set A by $|A|$.

A *finite automaton* (in short, an *automaton*) $\mathcal{A} = (S, X, \delta)$ consists of the following data: (1) S is a nonempty finite set, called a *state set.* (2) X is a nonempty finite alphabet. (3) δ is a function, called a *state transition function*, of $S \times X$ into S.

The state transition function δ can be extended to the function of $S \times X^*$ into S as follows: (1) $\delta(s, \epsilon) = s$ for any $s \in S$. (2) $\delta(s, au) = \delta(\delta(s, a), u)$ for any $s \in S, a \in X$ and $u \in X^*$.

Let $\mathcal{A} = (S, X, \delta)$ be an automaton, let $s \in S$ and let $u \in X^*$.. In what follows, we will write $su^{\mathcal{A}}$ instead of $\delta(s, u)$.

A *finite recognizer* $\mathcal{A} = (S, X, \delta, s_0, F)$ consists of the following data: (1) The triple (S, X, δ) constitutes a finite automaton. (2) $s_0 \in S$ is called the *initial state.* (3) $F \subseteq S$ is called the *set of final states.*

Let $\mathcal{A} = (S, X, \delta, s_0, F)$ be a finite recognizer. Then the language $\mathcal{T}(\mathcal{A}) = \{u \in X^* \mid \delta(s_0, u) \in F\}$ is called the *language accepted by* $\mathcal{A}$.

Let $L \subseteq X^*$. Then L is said to be *regular* if L is accepted by a finite recognizer.

2 Deterministic Directable Automata

First, we define a directable automaton.

Definition 1 An automaton $\mathcal{A} = (S, X, \delta)$ is said to be *directable* if the following condition is satisfied: There exists $w \in X^*$ such that $sw^{\mathcal{A}} = tw^{\mathcal{A}}$ for any $s, t \in S$.

In the above definition, a word $w \in X^*$ is called a *directing word* of $\mathcal{A}$. Then we have:

Fact *Let* $\mathcal{A} = (S, X, \delta)$ *be an automaton. Then* $\mathcal{A}$ *is directable if and only if for any* $s, t \in S$*, there exists* $u \in X^*$ *such that* $su^{\mathcal{A}} = tu^{\mathcal{A}}$.

Proposition 1 *Assume that* $\mathcal{A} = (S, X, \delta)$ *is a directable automata. Then the set of directing words* $\mathrm{D}(\mathcal{A})$ *of* $\mathcal{A}$ *is a regular language.*

Proof To prove the proposition, it is enough to provide the recognizer $\mathcal{B} = (P(S), X, \gamma, S, G)$ where $P(S) = \{T \mid T \subseteq S\}$, $G = (\{s\} \mid s \in S\}$ and $\gamma(T, a) = \bigcup_{t \in T} \delta(t, a)$ for any $a \in X$ and $T \in P(S)$. Then it can easily be verified that $\mathcal{T}(\mathcal{B}) = \mathrm{D}(\mathcal{A})$. Hence $\mathrm{D}(\mathcal{A})$ is regular.

Let $\mathcal{A} = (S, X, \delta)$ be a directable automaton. By $d(\mathcal{A})$, we denote the value $min\{|w| \mid w \in \mathrm{D}(\mathcal{A})\}$. Moreover, $d(n)$ denotes the value $max\{d(\mathcal{A}) \mid \mathcal{A} = (S, X, \delta)$ is a directable automaton with n states$\}$. In the definition of $d(n)$, X ranges over all finite nonempty alphabets.

In [2], Černý conjectured the following.

Conjecture *For any* $n \geq 1, d(n) = (n-1)^2$.

Actually, Černý proved only that $(n-1)^2 \leq d(n)$ using the following directable automaton.

$\mathcal{A} = (\{1, \ldots, n\}, \{a, b\}, \delta)$ where $\delta(i, a) = i + 1$ if $i = 1, 2, \ldots, n-1$ and $\delta(n, a) = 1$, $\delta(n-1, b) = n$ and $\delta(i, b) = i$ if $i = 1, 2, \ldots, n$ and $i \neq n-1$.

Then it can be shown that the word $b(a^{n-1}b)^{n-2}$ is a shortest directing word of $\mathcal{A}$ and hence $(n-1)^2 \leq d(n)$.

However, the above problem is still open and at present we have only the following result:

Proposition 2 *For any $n \geq 1$, we have $(n-1)^2 \leq d(n) \leq 1 + (n-2)\binom{n}{2}$.*

Proof The lower bound is obvious from the above result. As for the upper bound, it is enough to consider only the case $n \geq 3$. Let $\mathcal{A} = (S, X, \delta)$ be a directable automaton with $|S| = n$. Since $\mathcal{A}$ is a directable automaton, there exist $s_0, s_1 \in S, s_0 \neq s_1$ and $a \in X$ such that $s_0 a^{\mathcal{A}} = s_1 a^{\mathcal{A}}$. Hence $|Sa^{\mathcal{A}}| < |S|$. Let $s, t \in Sa^{\mathcal{A}}, s \neq t$. Moreover, let $u \in X^*$ be one of the shortest words such that $su^{\mathcal{A}} = tu^{\mathcal{A}}$. Suppose $|u| \geq \binom{n}{2} + 1$. Then we have $u = xvy, v, xy \in X^+$ such that $\{s,t\}x^{\mathcal{A}} = \{s,t\}(xv)^{\mathcal{A}}$. This implies that $s(xy)^{\mathcal{A}} = t(xy)^{\mathcal{A}}$. This contradicts the assumption that u is one of the shortest words. Hence $|u| \leq \binom{n}{2}$. Now consider $s', t' \in S(au)^{\mathcal{A}}, s' \neq t'$. In the same way as the above, we can find $u' \in X^*, |u'| \leq \binom{n}{2}$ such that $s'u'^{\mathcal{A}} = t'u'^{\mathcal{A}}$. Using the same technique, we can find a directing word $w = auu' \cdots$ of $\mathcal{A}$ with $|w| \leq 1 + (n-2)\binom{n}{2}$.

In [9] and [10], J.-E. Pin gave more precise evaluations.

We can consider a similar problem for some subclasses of directable automata. For instance, an automaton $\mathcal{A} = (S, X, \delta)$ is said to be *commutative* if $s(uv)^{\mathcal{A}} = s(vu)^{\mathcal{A}}$ holds for any $s \in S$ and any $u, v \in X^*$. By $d_{com}(n)$, we denote the value $max\{d(\mathcal{A}) \mid \mathcal{A} = (S, X, \delta)$ is commutative and directable, and $|S| = n\}$. In the definition of $d_{com}(n)$, X ranges over all finite nonempty alphabets. The following result is due to [11] and [12].

Proposition 3 *For any $n \geq 1$, we have $d_{com}(n) = n - 1$.*

Proof Let $\mathcal{A} = (S, X, \delta)$ be a commutative directable automaton with $|S| = n$. Remark that $Su^{\mathcal{A}} = \bigcup_{s \in S} su^{\mathcal{A}} \supseteq S(vu)^{\mathcal{A}} = S(uv)^{\mathcal{A}}$ for any $u, v \in X^*$. Let $w = a_1 a_2 \cdots a_r$ be one of the shortest directing words of $\mathcal{A}$. Thus we have $Sa_1^{\mathcal{A}} \supseteq S(a_1a_2)^{\mathcal{A}} \supseteq S(a_1a_2a_3)^{\mathcal{A}} \supseteq \cdots \supseteq S(a_1a_2a_3 \cdots a_r)^{\mathcal{A}}$. Notice that every inclusion is proper because the word w is one of the shotest directing words. Hence $r \leq n-1$. On the other hand, consider the following commutative directable automaton $\mathcal{B} = (\{1, 2, \ldots, n\}, X, \gamma)$ where $\gamma(i, a) = i + 1$ for any $a \in X$ and any $i = 1, 2, \ldots, n-1$ and $\gamma(n, a) = n$ for any $a \in X$. Then $d(\mathcal{B}) = n - 1$. Consequently, $d_{com}(n) = n - 1$.

It is an interesting question to ask whether or not a given automaton is directable. For this purpose, we introduce the notion of a merged word of a language.

Let $L \subseteq X^*$ be a language over X. Then $w \in X^*$ is called a *merged word* of L if any word in L is a subword of w, i.e. $X^*uX^* \cap \{w\} \neq \emptyset$ for any $u \in L$. In [8], the following proposition is given:

Proposition 4 *There exists a merged word $w \in X^+$ of $X^{d(n)}$ such that $|w| \leq |X|^{d(n)} + d(n) - 1$.*

Using the word w, we can check whether a given automaton $\mathcal{A} = (S, X, \delta)$ with $|S| = n$ is directable, i.e. $\mathcal{A}$ is directable if and only if $Sw^{\mathcal{A}}$ is a singleton set.

However, the above algorithm is not effective. To provide an effective algorithm to decide whether a given automaton is directable, we introduce the following relations ρ and ρ_i on S.

$\forall s, t \in S, (s, t) \in \rho \Longleftrightarrow \exists u \in X^*, su^{\mathcal{A}} = tu^{\mathcal{A}}$.

(1) $\forall s, t \in S, (s, t) \in \rho_0 \Longleftrightarrow s = t$.
(2) $\forall i \geq 1, \forall s, t \in S, (s, t) \in \rho_i \Longleftrightarrow \exists a \in X, (sa^{\mathcal{A}}, ta^{\mathcal{A}}) \in \rho_{i-1}$.

Then we have:

Proposition 5 $\rho = \bigcup_{i=0}^{\binom{n}{2}} \rho_i$. *Moreover, $\mathcal{A}$ is directable if and only if $(s, t) \in \rho$ for any $s, t \in S$.*

Using the above proposition, B. Imreh and M. Steinby ([5]) provided an algorithm to decide whether or not any given automaton $\mathcal{A} = (S, X, \delta)$ is directable with the time bound of $\mathcal{O}(m \cdot n^2)$ where $m = |X|$ and $n = |S|$.

3 Nondeterministic Directable Automata

A *nondeterministic automaton* $\mathcal{A} = (S, X, \delta)$ consists of the following data: (1) S, X are the same materials as in the definition of finite automata. (2) δ is a relation such that $\delta(s, a) \subseteq S$ for any $s \in S$ and any $a \in X \cup \{\epsilon\}$.

As in the case of finite automata, δ can be extended to the following relation in a natural way, i.e. $\delta(s, au) = \bigcup_{t \in \delta(s,a)} \delta(t, u)$ for any $s \in S$, any $u \in X^*$ and any $a \in X \cup \{\epsilon\}$. In what follows, we will write $su^{\mathcal{A}}$ instead of $\delta(s, u)$ as in the case of finite automata.

Now we will deal with nondeterministic directable automata and their related languages. For nondeterministic automata, the directability can be defined in several ways. In each case, the directing words constitute a regular language. We will consider six classes of regular languages with respect to the different definitions of directability.

Let $\mathcal{A} = (S, X, \delta)$ be a nondeterministic automaton. In [6], the notion of directing words of $\mathcal{A}$ is given. In the definition, $Sw^{\mathcal{A}}$ denotes $\bigcup_{s \in S} sw^{\mathcal{A}}$ for $w \in X^*$.

Definition 2 (1) A word $w \in X^*$ is D_1*-directing* if $sw^{\mathcal{A}} \neq \emptyset$ for any $s \in S$ and $|Sw^{\mathcal{A}}| = 1$. (2) A word $w \in X^*$ is D_2*-directing* if $sw^{\mathcal{A}} = Sw^{\mathcal{A}}$ for any $s \in S$. (3) A word $w \in X^*$ is D_3*-directing* if $\bigcap_{s \in S} sw^{\mathcal{A}} \neq \emptyset$.

Definition 3 Let $i = 1, 2, 3$. Then $\mathcal{A}$ is called a D_i*-directable automaton* if the set of D_i-directing words is not empty.

Let $\mathcal{A} = (S, X, \delta)$ be a nondeterministic automaton. Then, for any $i = 1, 2, 3$, $\mathrm{D}_i(\mathcal{A})$ denotes the set of all D_i-directing words. Then we have:

Proposition 6 *For any* $i = 1, 2, 3, \mathrm{D}_i(\mathcal{A})$ *is a regular language.*

Proof Let $\mathcal{A} = (S, X, \delta)$ be a nondeterministic automaton and let $S = \{s_0, s_1, s_2, \ldots, s_r\}, r \geq 0$. For any $i = 1, 2, 3$, we define a finite recognizer $\mathcal{A}_i = (\{T \mid T \subseteq S\}, X, \delta_r, (\{s_0\}, \{s_1\}, \ldots, \{s_r\}), F_i)$ as follows: (1) $(T_0, T_1, \ldots, T_r)a^{\mathcal{A}_i} = \delta_r((T_0, T_1, \ldots, T_r), a) = (T_0 a^{\mathcal{A}}, T_1 a^{\mathcal{A}}, \ldots, T_r a^{\mathcal{A}})$ for any $a \in X$ and any $T_i \subseteq S, i = 0, 1, \ldots, r$. (2) $F_1 = \{(\{t\}, \{t\}, \ldots, \{t\}) \mid t \in S\}$, $F_2 = \{(T, T, \ldots, T) \mid T \subseteq S\}$ and $F_3 = \{(T_0, T_1, \ldots, T_r) \mid T_i \subseteq S, i = 0, 1, \ldots, r, \bigcap_{i=0}^{r} T_i \neq \emptyset\}$.

Then it is obvious that $\mathrm{D}_i(\mathcal{A}) = \mathcal{T}(\mathcal{A}_i)$ for any $i = 1, 2, 3$. Hence $\mathrm{D}_i(\mathcal{A})$ is a regular language for any $i = 1, 2, 3$.

A nondeterministic automaton $\mathcal{A} = (S, X, \delta)$ is said to be *complete* if $sa^{\mathcal{A}} \neq \emptyset$ for any $s \in S$ and any $a \in X$. As for the D_1-directability of a complete nondeterministic automaton, Burkhard introduced it in [1]. We will investigate the classes of languages consisting of D_1-, D_2- and D_3-directing words of nondeterministic automata and complete nondeterministic automata.

The classes of D_i-directable nondeterministic automata and complete nondeterministic automata are denoted by $\mathbf{Dir}(i)$ and $\mathbf{CDir}(i)$, respectively. Let X be an alphabet. For $i = 1, 2, 3$, we define the following classes of languages:

(1) $\mathcal{L}^X_{\mathrm{ND}(i)} = \{\mathrm{D}_i(\mathcal{A}) \mid \mathcal{A} = (S, X, \delta) \in \mathbf{Dir}(i)\}$. (2) $\mathcal{L}^X_{\mathrm{CND}(i)} = \{\mathrm{D}_i(\mathcal{A}) \mid \mathcal{A} = (S, X, \delta) \in \mathbf{CDir}(i)\}$.

Let $\mathbf{D}$ be the class of deterministic directable automata. For $\mathcal{A} \in \mathbf{D}$, $\mathrm{D}(\mathcal{A})$ denotes the set of all directing words of $\mathcal{A}$. Then we can define the class, i.e. $\mathcal{L}^X_{\mathrm{D}} = \{\mathrm{D}(\mathcal{A}) \mid \mathcal{A} = (S, X, \delta) \in \mathbf{D}\}$.

Then, by Propsition 1 and Proposition 6, all the above classes are subclasses of regular languages. Figure 1 depicts the inclusion relations among such 7 classes. In [3], the inclusion relations among more classes are provided.

We will consider the shortest directing words of nondeterministic automata.

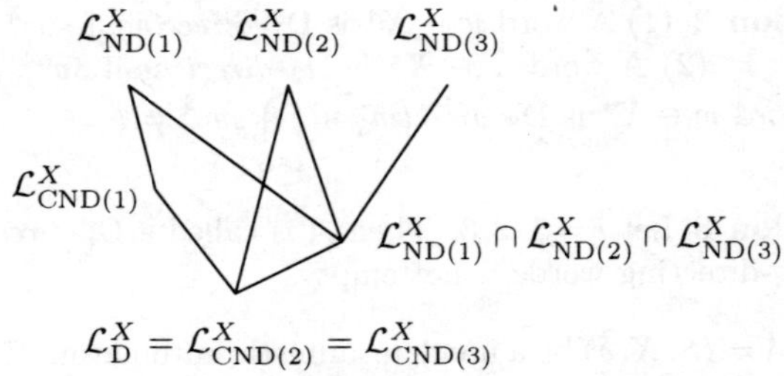

Figure 1: Inclusion relations

Let $i = 1, 2, 3$ and let $\mathcal{A} = (S, X, \delta)$ be a nondeterministic automaton. Then $d_i(\mathcal{A})$ denotes the value $min\{|u| \mid u \in \mathrm{D}_i(\mathcal{A})\}$. For any positive integer $n \geq 1$, $d_i(n)$ denotes the value $max\{d_i(\mathcal{A}) \mid \mathcal{A} = (S, X, \delta) : \mathcal{A} \in \mathbf{Dir}(i) \text{ and } |S| = n\}$. Moreover, $cd_i(n)$ denotes the value $max\{d_i(\mathcal{A}) \mid \mathcal{A} = (S, X, \delta) : \mathcal{A} \in \mathbf{CDir}(i)$ and $|S| = n\}$. Notice that in the definitions of $d_i(n)$ and $cd_i(n)$, X ranges over all finite nonempty alphabets.

In [1], Burkhard determined the value $cd_1(n)$ as follows:

Proposition 7 *Let $n \geq 1$. Then $cd_1(n) = 2^n - n - 1$.*

Proof First, we show that $cd_1(n) \leq 2^n - n - 1$. Let $\mathcal{A} = (S, X, \delta)$ be a complete nondeterministic automaton with $|S| = n$ and let $w = a_1 a_2 \cdots a_{r-1} a \in \mathrm{D}_1(\mathcal{A})$ where $a_1, a_2, \ldots, a_{r-1}, a \in X$ and $r = |w| = d_1(\mathcal{A})$. Suppose that $r = |w| > 2^n - n - 1$. Notice that $2^n - n - 2 = |\{T \subset S \mid |T| \geq 2\}|$. Let $T_i = S(a_1 a_2 \cdots a_i)^{\mathcal{A}}$ where $i = 1, 2, \ldots, r-1$, Then there exist $i, j = 1, 2, \ldots, r-1$ such that $i < j$ and $T_i = T_j$. In this case, $a_1 a_2 \cdots a_i a_{j+1} \cdots a_r a \in \mathrm{D}_1(\mathcal{A})$. This contradicts the assumption that $r = d_1(\mathcal{A})$. Hence $cd_1(n) \leq 2^n - n - 1$.

Now we show that $2^n - n - 1 \leq cd_1(n)$. To prove $2^n - n - 1 \leq cd_1(n)$, it is enough to construct a complete nondeterministic automaton $\mathcal{A} = (S, X, \delta)$ such that $|S| = n$ and $d_1(\mathcal{A}) = 2^n - n - 1$. Let S be a finite set with $|S| = n$. Moreover, let $\{T_1, T_2, \ldots, T_r\} = \{T \subset S \mid |T| \geq 2\}$. Furtheremore, we assume that $|T_1| \geq |T_2| \geq \cdots \geq |T_r|$ and $T_r = \{s_1, s_2\}$. Notice that $r = 2^n - n - 2$. Now we construct the following nondeterministic automaton $\mathcal{A} = (S, X, \delta)$: (1) $X = \{a_1, a_2, \ldots, a_r, z\}$. (2) $sa_1^{\mathcal{A}} = T_1$ for any $s \in S$. (3) For any $i = 1, 2, \ldots, r-1$, $sa_{i+1}^{\mathcal{A}} = T_{i+1}$ if $s \in T_i$ and $sa_{i+1}^{\mathcal{A}} = S$, otherwise. (4) $s_1 z^{\mathcal{A}} = s_2 z^{\mathcal{A}} = \{s_1\}$ and $sz^{\mathcal{A}} = S$ if $s \neq s_1, s_2$.

Then $\mathcal{A}$ is complete and $S(a_1 a_2 \cdots a_r z)^{\mathcal{A}} = Sa_1^{\mathcal{A}}(a_2 \cdots a_r z)^{\mathcal{A}} = T_1(a_2 \cdots a_r z)^{\mathcal{A}} = T_1 a_2^{\mathcal{A}}(a_3 \cdots a_r z)^{\mathcal{A}} = T_2(a_3 \cdots a_r z)^{\mathcal{A}} = \cdots = T_r z^{\mathcal{A}} = \{s_1, s_2\} z^{\mathcal{A}} = \{s_1\}$. Hence $a_1 a_2 \cdots a_r z \in \mathrm{D}_1(\mathcal{A})$ and $|a_1 a_2 \cdots a_r z| = r + 1 = 2^n - n - 1$. Suppose that there exists $w \in \mathrm{D}_1(\mathcal{A})$ with $d_1(\mathcal{A}) = |w| < 2^n - n - 1$. Let

$w = a_i w'$ where $w' \in X^*$ and $i = 2, 3, \ldots, r$. Since $S \supset T_{i-1}$, we have $Sa_i = S$, $S(a_i w')^{\mathcal{A}} = Sw'^{\mathcal{A}}$ and $|Sw'^{\mathcal{A}}| = 1$. This implies that $w' \in \mathrm{D}_1(\mathcal{A})$. This contradicts the minimality of $|w|$. For the case that $w = zw'$ where $w' \in X^*$, we have $S(zw')^{\mathcal{A}} = Sz^{\mathcal{A}}w'^{\mathcal{A}} = Sw'^{\mathcal{A}}$ because $S \supset \{s_1, s_2\}$. Hence we encounter the same contradiction. Assume $w = a_1a_2 \cdots a_i w'$ for some $i = 1, 2, \ldots, r$ and $w' \in X^*$. Obviously, $w' \in X^+$ and hence $i \leq r - 1$. Suppose $w' = a_j v$ where $j \neq i + 1$ and $v \in X^*$. Let $j < i + 1$. Remark that $w = a_1a_2 \cdots a_j \cdots a_i a_j v$. Notice that $S(a_1a_2 \cdots a_j \cdots a_i a_j)^{\mathcal{A}} = T_j$ or $S(a_1a_2 \cdots a_j \cdots a_i a_j)^{\mathcal{A}} = S$. In the former case, since $S(a_1a_2 \cdots a_j)^{\mathcal{A}} = T_j$, we have $S(a_1a_2 \cdots a_j v)^{\mathcal{A}} = Sw^{\mathcal{A}}$ and $a_1a_2 \cdots a_j v \in \mathrm{D}_1(\mathcal{A})$, which contradicts the minimality of $|w|$. In the latter case, $Sw^{\mathcal{A}} = Sv^{\mathcal{A}}$ and $|Sv^{\mathcal{A}}| = 1$. Therefore, $v \in \mathrm{D}_1(\mathcal{A})$, which is a contradiction. Suppose $j > i + 1$. Since $i < j - 1$, $T_i \setminus T_{j-1} \neq \emptyset$. Therefore, $S(a_1a_2 \cdots a_i a_j)^{\mathcal{A}} = T_i a_j^{\mathcal{A}} = S$. Consequently, $Sv^{\mathcal{A}} = Sw^{\mathcal{A}}$ and $v \in \mathrm{D}_1(\mathcal{A})$, which contradicts the minimality of $|w|$. Now let $w' = zv$ where $v \in X^*$. Since $w = a_1a_2 \cdots a_i zv$, $Sw^{\mathcal{A}} = S(a_1a_2 \cdots a_i zv)^{\mathcal{A}} = T_i z^{\mathcal{A}} v^{\mathcal{A}}$. Notice that $T_i \neq T_r$. Hence $T_i \setminus \{s_1, s_2\} \neq \emptyset$. Therefore, $Sw^{\mathcal{A}} = S(a_1a_2 \cdots a_i zv)^{\mathcal{A}} = T_i z^{\mathcal{A}} v^{\mathcal{A}} = Sv^{\mathcal{A}}$ and $v \in \mathrm{D}_1(\mathcal{A})$, which contradicts the minimality of $|w|$. This means that there is no $w \in \mathrm{D}_1(\mathcal{A})$ with $|w| < 2^n - n - 1$. Hence $d_1(\mathcal{A}) = 2^n - n - 1$. This completes the proof of the proposition.

For $d_1(n)$, we have the following new result.

Proposition 8 *Let $n \geq 2$. Then $2^n - n \leq d_1(n) \leq \sum_{k=2}^{n} \binom{n}{k}(2^k - 1)$. Notice that $d_1(1) = 0$ and $d_1(2) = 3$.*

Proof Let $n \geq 2$. First, we show that $d_1(n) \leq \sum_{k=2}^{n} \binom{n}{k}(2^k - 1)$. Let $\mathcal{A} = (S, X, \delta)$ be a D_1-directable automaton with n states and let $w = a_1a_2 \cdots a_r \in \mathrm{D}_1(\mathcal{A})$ such that $a_i \in X, i = 1, 2, \ldots, r, r \geq 1$ and $|w| = r = d_1(\mathcal{A})$. Since $w \in \mathrm{D}_1(\mathcal{A})$, there exists $s_0 \in S$ such that $sw^{\mathcal{A}} = \{s_0\}$ for any $s \in S$. For any $i = 1, 2, \ldots, r$, we define the set S_i and T_i as follows: (1) $S_i = S(a_1a_2 \cdots a_i)^{\mathcal{A}}$. (2) $T_i = \{t \in S_i \mid t(a_{i+1}a_{i+2} \cdots a_r)^{\mathcal{A}} = \{s_0\}\}$.

Let $s \in S$ and let $i = 1, 2, \ldots, r$. Since $s(a_1a_2 \cdots a_i a_{i+1} \cdots a_r)^{\mathcal{A}} = (s(a_1a_2 \cdots a_i)^{\mathcal{A}})(a_{i+1} \cdots a_r)^{\mathcal{A}} = \{s_0\}$, we have $s(a_1a_2 \cdots a_i)^{\mathcal{A}} \cap T_i \neq \emptyset$. Let $S = S_0 = T_0$. Consider the set $\{(S_i, T_i) \mid i = 0, 1, 2, \ldots, r - 1\}$. It is obvious that $S_i \neq \emptyset$ for any $i = 0, 1, \ldots, r - 1$. It is also obvious that $|S_0| \neq 1$. Suppose that $|S_i| = 1$ for some $i = 1, 2, \ldots, r - 1$. Then $S_i = T_i = \{t\}$ for some $t \in S$. By the definition of T_i, this means that $a_1a_2 \cdots a_i \in \mathrm{D}_1(\mathcal{A})$, which contradicts the minimality of $|w|$. Therefore, $|S_i| \neq 1$ for any $i = 1, 2, \ldots, n$. Hence the set $\{(S_i, T_i) \mid i = 0, 1, 2, \ldots, r - 1\}$ does not contain any $(\{s\}, \{s\})$ with $s_0 \neq s \in S$.

Now assume that $(S_i, T_i) = (S_j, T_j)$ for some $i, j = 1, 2, \ldots, r-1, i < j$. Then it can be seen that $a_1a_2 \cdots a_i a_{j+1}a_{j+2} \cdots a_r \in \mathrm{D}_1(\mathcal{A})$, which contradicts the minimality of $|w|$. Hence all $(S_i, T_i), i = 0, 1, 2, \ldots, r-1$, are distinct. Therefore, $|\{(S_i, T_i) \mid i = 0, 1, 2, \ldots, r-1\}| \leq \sum_{k=2}^{n} \binom{n}{k}(2^k - 1)$ and hence $r \leq \sum_{k=2}^{n} \binom{n}{k}(2^k - 1)$.

We will show that $2^n - n \leq d_1(n)$. It is obvious that $d_1(2) \geq 2$. Let $n \geq 3$. We will construct a D_1-directable automaton $\mathcal{A} = (S, X, \delta)$ such that $|S| = n$ and $d_1(\mathcal{A}) = 2^n - n$. Let S be a finite set with $|S| = n$ and let $\{T_1, T_2, \ldots, T_r\} = \{T \subset S \mid |T| \geq 2\}$. Notice that $r = 2^n - n - 2$. Moreover, we assume that $|T_1| \geq |T_2| \geq \cdots \geq |T_r|$, $\{s_0\} = S \setminus T_1$ and $T_r = \{s_1, s_2\}$. Now we construct the following nondeterministic automaton $\mathcal{A} = (S, X, \delta)$: (1) $X = \{a_1, a_2, \ldots, a_r, b\}$. (2) For any $i = 1, 2, \ldots, r-1$, $sa_i^{\mathcal{A}} = T_{i+1}$ if $s \in T_i$ and $sa_i^{\mathcal{A}} = S$, otherwise. (3) $s_1 a_r^{\mathcal{A}} = s_2 a_r^{\mathcal{A}} = \{s_1\}$ and $sa_r^{\mathcal{A}} = S$ if $s \in S \setminus \{s_1, s_2\}$. (4) $s_0 b^{\mathcal{A}} = \emptyset$ and $sb^{\mathcal{A}} = T_1$ for any $s \in S \setminus \{s_0\}$.

Let $s \in S$ and let $i = 1, 2, \ldots, r$. Notice that $s(a_i b a_1 a_2 \cdots a_r)^{\mathcal{A}} = \{s_1\}$ and hence $a_i b a_1 a_2 \cdots a_r \in \mathrm{D}_1(\mathcal{A})$. Moreover, since $s_0 b^{\mathcal{A}} = \emptyset$, we have $bX^* \cap \mathrm{D}_1(\mathcal{A}) = \emptyset$. Let $i, j = 1, 2, \ldots, r$. Then $S(a_i a_j)^{\mathcal{A}} = S$. On the other hand, $s(a_i b)^{\mathcal{A}} = T_1$ for any $s \in S$. This means that $u \in a_i b X^*$ if u is a shortest D_1-directing word of $\mathcal{A}$. Let $i = 1, 2, \ldots, r-1$. Then $T_i(a_i a_j)^{\mathcal{A}} = T_{i+1} a_j^{\mathcal{A}} = S$ if $j > i+1$ and $T_i(a_i a_j)^{\mathcal{A}} = T_{i+1} a_j^{\mathcal{A}} \supseteq T_{j+1}$ if $j \leq i$. Notice that in the latter case $j+1 \leq i+1$. This implies that u is not a shortest D_1-directing word of $\mathcal{A}$ if $u \in X^* a_i a_j X^*$ where $j \neq i+1$. Moreover, since $Sb^{\mathcal{A}} = T_1$, u is not a shortest D_1-directing word of $\mathcal{A}$ if $u \in XX^+ bX^*$. Consequently, $a_i b a_1 a_2 \cdots a_r$ is a shortest D_1-directing word of $\mathcal{A}$, i.e. $d_1(\mathcal{A}) = r + 2 = 2^n - n$. Hence we have $2^n - n \leq d_1(n)$.

Finally, we compute $d_1(1)$ and $d_1(2)$. It is obvious that $d_1(1) = 0$. Consider the following nondeterministic automaton $\mathcal{A} = (\{1, 2\}, \{a, b, c\}, \delta)$: (1) $1a^{\mathcal{A}} = \{1, 2\}$ and $2a^{\mathcal{A}} = \{2\}$. (2) $1b^{\mathcal{A}} = \emptyset$ and $2b^{\mathcal{A}} = \{1, 2\}$. (3) $1c^{\mathcal{A}} = \{1\}$ and $2c^{\mathcal{A}} = \emptyset$.

Then abc is a shortest D_1-directing word of $\mathcal{A}$. Since $d_1(2) \leq 2^2 - 1 = 3$, we have $d_1(2) = 3$.

Now we consider the value $d_3(n)$. Before dealing with the value $d_3(n)$, we define a nondeterministic automaton of partial function type.

A nondeterministic automaton $\mathcal{A} = (S, X, \delta)$ is said to be *of partial function type* if $|sa^{\mathcal{A}}| \leq 1$ for any $s \in S$ and any $a \in X$. Then we have:

Remark 1 Let $\mathcal{A}$ be a nondeterministic automaton of partial function type. Then $\mathrm{D}_3(\mathcal{A}) = \mathrm{D}_1(\mathcal{A})$.

Let $\mathcal{A} = (S, X, \delta)$ be a D_3-directable automaton of partial function type. Consider the following procedure $\mathcal{P}$: Let $u \in \mathrm{D}_3(\mathcal{A})$. Assume that $u = u_1 u_2 u_3$ where $u_1, u_3 \in X^*, u_2 \in X^+$ and $Su_1{}^{\mathcal{A}} = S(u_1 u_2)^{\mathcal{A}}$. Then procedure $\mathcal{P}$ can be applied as $u \Rightarrow^{\mathcal{P}} u_1 u_3$.

Then we have the following result.

Lemma 1 *In the above procedure, we have* $u_1u_3 \in \mathrm{D}_3(\mathcal{A})$.

Proof Let $\mathcal{A} = (S, X, \delta)$ be a nondeterministic automaton of partial function type. Moreover, let $u = u_1u_2u_3$ where $u_1, u_3 \in X^*, u_2 \in X^+$ and $Su_1{}^{\mathcal{A}} = S(u_1u_2)^{\mathcal{A}}$. Since $u \in \mathrm{D}_3(\mathcal{A})$, there exists $s_0 \in S$ such that $su^{\mathcal{A}} = \{s_0\}$ for any $s \in S$. From the assumptions that $Su_1{}^{\mathcal{A}} = S(u_1u_2)^{\mathcal{A}}$ and $\mathcal{A}$ is a nondeterministic automaton of partial function type, it follows that $su^{\mathcal{A}} = s(u_1u_2u_3)^{\mathcal{A}} = s(u_1u_3)^{\mathcal{A}} = \{s_0\}$ for any $s \in S$. By Remark 1, this means that $u_1u_3 \in \mathrm{D}_3(\mathcal{A})$.

Let $\mathcal{A} = (S, X, \delta)$ be a D_3-directable automaton of partial function type and let $a_1a_2 \cdots a_r \in \mathrm{D}_3(\mathcal{A})$ such that $sa_1{}^{\mathcal{A}} = ta_1{}^{\mathcal{A}}$ for some $s, t \in S, s \neq t$.

Assume that $v \in \mathrm{D}_3(\mathcal{A}), v = v_1v_2v_3, v_1, v_3 \in X^*, v_2 \in X^+$, $|Sv_1{}^{\mathcal{A}}| = |S(v_1v_2)^{\mathcal{A}}|$ and $\{s,t\} \subseteq Sv_1{}^{\mathcal{A}}$. Then procedure $\mathcal{Q}_{(s,t)}$ can be applied as $v \Rightarrow^{\mathcal{Q}_{(s,t)}} v_1a_1a_2 \cdots a_r$.

Then we have the following results.

Lemma 2 *In the above procedure, we have* $v_1a_1a_2 \cdots a_r \in \mathrm{D}_3(\mathcal{A})$ *and* $|Sv_1{}^{\mathcal{A}}| > |Sv_1a_1{}^{\mathcal{A}}|$.

Proof Let $s \in S$. Since $v = v_1v_2v_3 \in \mathrm{D}_3(\mathcal{A})$, we have $sv_1{}^{\mathcal{A}} \neq \emptyset$, actually $|sv_1{}^{\mathcal{A}}| = 1$. Notice that $\exists s_r \in S, \forall t \in S, t(a_1a_2 \cdots a_r)^{\mathcal{A}} = \{s_r\}$. Therefore, $s(v_1a_1a_2 \cdots a_r)^{\mathcal{A}} = (sv_1{}^{\mathcal{A}})(a_1a_2 \cdots a_r)^{\mathcal{A}} = \{s_r\}$ and hence $v_1a_1a_2 \cdots a_r \in \mathrm{D}_3(\mathcal{A})$. Since $\mathcal{A}$ is of partial function type and $\{s,t\} \subseteq Sv_1{}^{\mathcal{A}}$, $|Sv_1{}^{\mathcal{A}}| \geq |Sv_1a_1{}^{\mathcal{A}}| + 1$. This completes the proof of the lemma.

Lemma 3 *Let* $\mathcal{A} = (S, X, \delta)$ *be a* D_3*-directable automaton such that* $|S| = n$ *and* $d_3(\mathcal{A}) = d_3(n)$. *Then there exists a nondeterministic automaton* $\mathcal{B} = (S, Y, \gamma)$ *of partial function type such that* $d_3(\mathcal{B}) = d_3(n)$.

Proof Let $u = a_1a_2 \cdots a_r \in \mathrm{D}_3(\mathcal{A})$ with $|u| = d_3(\mathcal{A})$. Since $u \in \mathrm{D}_3(\mathcal{A})$, there are $s_r \in S$ and a sequence of partial functions of S into S, $\rho_1, \rho_2, \ldots, \rho_r$ such that $s(a_1a_2 \cdots a_i)^{\mathcal{A}} \supseteq \rho_i(\rho_{i-1}(\cdots(\rho_1(s))\cdots))$ for any $s \in S$ and any $i = 1, 2, \ldots, r$. Furthermore, $\rho_r(\rho_{r-1}(\cdots(\rho_1(s))\cdots)) = \{s_r\}$ for any $s \in S$. Now we define the automaton of partial function type $\mathcal{B} = (S, Y, \gamma)$ as follows: (1) $Y = \{b_i \mid i = 1, 2, \ldots, r\}$. Remark that $b_1, b_2, \ldots, b_r$ are distinct symbols. (2) $sb_i{}^{\mathcal{B}} = \rho_i(s)$ for any $s \in S$ and any $i = 1, 2, \ldots, r$.

Then $\mathcal{B}$ is a nondeterministic automaton of partial function type. Moreover, it is obvious that $b_1b_2 \cdots b_r \in \mathrm{D}_3(\mathcal{B})$. Suppose that $b_{i_1}b_{i_2} \cdots b_{i_k} \in \mathrm{D}_3(\mathcal{B})$ where $i_1, i_2, \ldots, i_k \in \{1, 2, \ldots, r\}$. Then we have $a_{i_1}a_{i_2} \cdots a_{i_k} \in \mathrm{D}_3(\mathcal{A})$. Therefore, $k \geq r$ and $r = d_3(\mathcal{B})$. This completes the proof of the lemma.

We are now ready to determine an upper bound for $d_3(n)$.

Proposition 9 *For any* $n \geq 3$, $d_3(n) \leq \sum_{k=2}^{n-1} \binom{n}{k} - \sum_{k=0}^{n-2} \binom{n-2}{k} + n - 1$.

Proof By Lemma 3, there exists a nondeterministic automaton of partial function type $\mathcal{A} = (S, X, \delta)$ such that $|S| = n$ and $d_3(n) = d_3(\mathcal{A})$. Let $u = a_1 a_2 \cdots a_r \in \mathrm{D}_3(\mathcal{A})$ with $r = d_3(n)$ and let $S_i = S(a_1 a_2 \cdots a_i)^{\mathcal{A}}$ for $i = 1, 2, \ldots, r$. Since $\mathcal{A}$ is of partial function type and $r = d_3(n) = d_3(\mathcal{A})$, $|S| > |S_1| \geq |S_2| \geq \cdots \geq |S_{r-1}| > |S_r| = 1$. Let $S_r = \{s_r\}$. By Lemma 1, $S, S_1, S_2, \ldots, S_{r-1}$ and S_r are distinct. Moreover, since $|S| > |S_1|$, there exist $s_0, s_1 \in S$ such that $s_0 \neq s_1$ and $s_0 {a_1}^{\mathcal{A}} = s_1 {a_1}^{\mathcal{A}}$. Therefore, we can apply procedure $\mathcal{Q}_{(s_0,s_1)}$ to $a_1 a_2 \cdots a_r$ if necessary and we can get $a_1 a_2 \cdots a_r \Rightarrow^{\mathcal{Q}_{(s_0,s_1)}} v_1 a_1 a_2 \cdots a_r$. Now we apply procedure $\mathcal{P}$ to $v_1 a_1 a_2 \cdots a_r$ as many times as possible until we cannot apply procedure $\mathcal{P}$ anymore. Hence we can obtain $w \in \mathrm{D}_3(\mathcal{A})$ with $|w| \leq 2^{|S|} - |S|$. Then we apply procedure $\mathcal{Q}_{(s_0,s_1)}$ to w. We will continue the same process until we cannot apply either procedure $\mathcal{P}$ nor $\mathcal{Q}_{(s_0,s_1)}$. Notice that this process will be terminated after a finite number of applications of procedures $\mathcal{P}$ and $\mathcal{Q}_{(s_0,s_1)}$. Let $w = c_1 c_2 \cdots c_s, c_i \in X, i = 1, 2, \ldots, s$ be the last D_3-directing word of $\mathcal{A}$ which was obtained by the above process. Let $T_i = S(c_1 c_2 \cdots c_i)^{\mathcal{A}}$ for any $i = 1, 2, \ldots, s$. Then $T_i \neq T_j$ for any $i, j = 1, 2, \ldots, s$ with $i < j$ and $\{T_1, T_2, \ldots, T_s\}$ contains at most $n-2$ elements $T_i, i = 1, 2, \ldots, s$ with $T_i \supseteq \{s_0, s_1\}$. Since $|\{T \subseteq S \mid \{s_0, s_1\} \subseteq T\}| = \sum_{k=0}^{n-2} \binom{n-2}{k}$ and by the above observation (including Lemma 2), we have $d_3(n) \leq \sum_{k=2}^{n-1} \binom{n}{k} - \sum_{k=0}^{n-2} \binom{n-2}{k} + n - 1$.

For the lower bound for $d_3(n)$, we have the following new result.

Proposition 10 *Let* $n \geq 3$. *Then* $d_3(n) \geq 2^m + 1$ *if* $n = 2m$ ($d_3(n) \geq 3 \cdot 2^{m-1} + 1$ *if* $n = 2m + 1$).

Proof Let $n \geq 3$ and let $S = \{1, 2, \ldots, n\}$. Moreover, let $S_1 = \{1, 2\}$, let $S_2 = \{3, 4\}, \ldots$, let $S_{m-1} = \{2m-3, 2m-2\}$ and let $S_m = \{2m-1, 2m\}$ if $n = 2m$ ($S_m = \{2m-1, 2m, 2m+1\}$ if $n = 2m+1$).

We define the following D_3-directable automaton $\mathcal{A} = (S, X, \delta)$:

(1) $\{T_1, T_2, \ldots, T_k\} = \{\{n_1, n_2, \ldots, n_m\} \mid (n_1, n_2, \ldots, n_m) \in S_1 \times S_2 \times \cdots \times S_m\}$ where $k = 2^m$ if $n = 2m$ ($k = 3 \cdot 2^{m-1}$ if $n = 2m+1$). (2) $T_1 = \{1, 3, 5, \ldots, 2m-1\}$. (3) $X = \{a, b_1, b_2, \ldots, b_{k-2}, b_{k-1}, c\}$. (4) $1a^{\mathcal{A}} = 2a^{\mathcal{A}} = \{1\}, 3a^{\mathcal{A}} = 4a^{\mathcal{A}} = \{3\}, \ldots, (2m-3)a^{\mathcal{A}} = (2m-2)a^{\mathcal{A}} = \{2m-3\}$ and $(2m-1)a^{\mathcal{A}} = (2m)a^{\mathcal{A}} = \{2m-1\}$ if $n = 2m$ ($(2m-1)a^{\mathcal{A}} = (2m)a^{\mathcal{A}} = (2m+1)a^{\mathcal{A}} = \{2m-1\}$ if $n = 2m+1$). (5) Let $i = 1, 2, \ldots, k-1$. By ρ_i, we denote a bijection of T_i onto T_{i+1}. Then $t{b_i}^{\mathcal{A}} = \rho_i(t)$ for any $t \in T_i$ and $t{b_i}^{\mathcal{A}} = \emptyset$, otherwise. (6) $tc^{\mathcal{A}} = \{1\}$ for any $t \in T_k$ and $tc^{\mathcal{A}} = \emptyset$, otherwise.

Then it can be easily verified that $ab_1b_2\cdots b_{k-1}c$ is a unique shortest D_3-directing word of $\mathcal{A}$. Therefore, $d_3(n) \geq 2^m + 1$ if $n = 2m$ ($d_3(n) \geq 3 \cdot 2^{m-1} + 1$ if $n = 2m + 1$).

Now we consider the values $cd_2(n)$ and $d_2(n)$. The lower bound is due to [1] and the upper bound is followed by [6].

Proposition 11 *For* $n \geq 2$, $2^n - n - 1 \leq cd_2(n) \leq d_2(n) < 1 + (2^n - 2)\binom{2^n}{2}$. *Remark that* $cd_2(1) = d_2(1) = 0$.

Finally, we provide a result on the value of $cd_3(n)$. The result is due to [2] and [6].

Proposition 12 *Let* $n \geq 1$. *Then* $(n-1)^2 \leq cd_3(n) \leq 1 + (n-2)\binom{n}{2}$.

4 Commutative Nondeterministic Directable Automata

In this section, we will deal with commutative nondeterministic automata and related languages alongside the same line as that of the previous section.

A nondeterministic automaton $\mathcal{A} = (S, X, \delta)$ is said to be *commutative* if $s(ab)^{\mathcal{A}} = s(ba)^{\mathcal{A}}$ holds for any $s \in S$ and any $a, b \in X$.

By $\mathcal{L}'^X_{\mathrm{D}}, \mathcal{L}'^X_{\mathrm{CND}(i)}$ and $\mathcal{L}'^X_{\mathrm{ND}(j)}, i, j = 1, 2, 3$, we denote the classes of regular languages of directing words of deterministic commutative automata, of D_i-directing words of complete commutative nondeterministic automata, and of D_j-directing words of commutative nondeterministic automata, respectively.

Then we have the following inclusion relations among these classes (see Figure 2).

$$\mathcal{L}'^X_{\mathrm{ND}(1)} = \mathcal{L}'^X_{\mathrm{ND}(3)}$$

$$\big|$$

$$\mathcal{L}'^X_{\mathrm{D}} = \mathcal{L}'^X_{\mathrm{ND}(2)} = \mathcal{L}'^X_{\mathrm{CND}(1)} = \mathcal{L}'^X_{\mathrm{CND}(2)} = \mathcal{L}'^X_{\mathrm{CND}(3)}$$

Figure 2: Commutative case

Now we will consider the shortest directing words of commutative nondeterministic automata. The results in this section are due to [4].

Let $i = 1, 2, 3$ and let $n \geq 1$. Then $cd_{com(i)}(n)$ denotes the value $max\{d_i(\mathcal{A}) \mid \mathcal{A} = (S, X, \delta)$ is commutative, $\mathcal{A} \in \mathbf{CDir}(i)$ and $|S| = n\}$.

Notice that in the definitions of $d_{com(i)}(n)$ and $cd_{com(i)}(n)$, X ranges over all finite nonempty alphabets.

Proposition 13 *For any* $n \geq 1, d_{com(1)}(n) = cd_{com(1)}(n) = n - 1$.

Proposition 14 *Let* $n \geq 2$. *Then* $(n-1)^2 + 1 \leq cd_{com(2)}(n) = d_{com(2)}(n) \leq 2^n - 2$. *For* $n = 1$, $cd_{com(2)}(1) = d_{com(2)}(1) = 0$.

Proposition 15 *Let* $n \geq 2$. *Then* $n^2 - 3n + 3 \leq cd_{com(3)}(n) = d_{com(3)}(n) \leq 1 + (n-2)\binom{n}{2}$. *For* $n = 1, cd_{com(3)}(1) = d_{com(3)}(1) = 0$.

As for more detailed information on deterministic and nondeterministic directable automata, refer to [7].

Acknowledgement The author would like to thank Dr. K. Tsuji and Dr. Cs. Imreh for their valuable comments.

References

[1] H.V. Burkhard, Zum Längenproblem homogener experimente an determinierten und nicht-deterministischen automaten, Elektronische Informationsverarbeitung und Kybernetik, EIK 12 (1976), 301-306.

[2] J. Černý, Poznámka k homogénym experimentom s konečinými automatami, Matematicko-fysikalny Časopis SAV 14 (1964), 208-215.

[3] B. Imreh and M. Ito, On some special classes of regular languages, in *Jewels are Forever* (edited by J. Karhumäki et al.) (1999) (Springer, New York), 25-34.

[4] B. Imreh, M. Ito and M. Steinby, On commutative directable nondeterministic automata, in *Grammars and Automata for Strings: From Mathematics and Computer Science to Biology, and Back* (edited by C. Martin-Vide et al.) (2003) (Taylor and Francis, London), 141-150.

[5] B. Imreh and M. Steinby, Some remarks on directable automata, Acta Cybernetica 12 (1995), 23-36.

[6] B. Imreh and M. Steinby, Directable nondeterministic automata, Acta Cybernetica 14 (1999), 105-115.

[7] M. Ito, *Algebraic Theory of Automata and Languages*, World Scientific (Singapore), 2004.

[8] M. Ito and J. Duske, On cofinal and definite automata, Acta Cybernetica 6 (1983), 181-189.

[9] J.-E. Pin, Sur les mots synchronisants dans un automata fini, Elektronische Informationsverarbeitung und Kybernetik, EIK 14 (1978), 297-303.

[10] J.-E. Pin, Sur un cas particulier de la conjecture de Cerny, Automata, Lecture Notes in Computer Science 62 (Springer) (1979), 345-352.

[11] I. Rystsov, Exact linear bound for the length of reset words in commutative automata, Publicationes Mathematicae of Debrecen 48 (1996), 405-409.

[12] I. Rystsov, Reset words for commutative and solvable automata, Theoretical Computer Science 172 (1997), 273-279

Worst-Case Redundancy of Solid Codes*

H. Jürgensen
Department of Computer Science
The University of Western Ontario
London, Ontario, Canada N6A 5B7
and
Institut für Informatik
Universität Potsdam
August-Bebel-Straße 89
14482 Potsdam, Germany

S. Konstantinidis
Department of Mathematics and Computing Science
Saint Mary's University
Halifax, Nova Scotia, Canada, B3H 3C3

Abstract

The quality of codes can be expressed in terms of their error-resistance and their redundancy. Solid codes are highly resistant to decoding errors resulting from incorrectly transmitted code words. We derive lower bounds on the redundancy of certain maximal solid codes.

1 Introduction

Solid codes, introduced in [5, 6] for information transmission in the presence of certain types of synchronization errors, are remarkably error-resistant in the following sense: in a received message every correctly transmitted code word will be decoded correctly. In a received message there may, of course, be parts that have been disturbed beyond repair or parts that, due to errors, happen to be code words, but not the ones originally sent; however, the decoding of those

*This research was supported by the Natural Sciences and Engineering Council of Canada, through Grants OGP000243 and OGP220259.

parts that have been transmitted correctly will not be affected by the errors in their vicinity when a solid code has been employed. Using this property as a definition, solid codes were re-introduced in [7] in the context of the study of disjunctive domains. First steps towards a systematic investigation of the error-detection capabilities of solid codes is presented in [3]. A survey of known results concerning solid codes, as of 1996, is available in [2].

When introducing error-handling capabilities like error-resistance, error-detection or error-correction, one pays a price in terms of transmission speed. For systematic block codes the speed is usually expressed as the ratio of the number of information bits over the code word length. For variable-length codes, one can express the transmission speed – or rather the transmission delay – in terms of the redundancy introduced by using the code.

Depending on the context, redundancy has been defined in several quite different ways including the following.

1. Given a probabilistic model for an information source, redundancy can be defined as the difference between the expected code word length and the entropy of the source.

2. Without a probabilistic model for the source, redundancy can be expressed as the difference between the maximal length of a code word in a given code and the optimal length of a code word given a fixed number of distinct signals.

In this paper – as in [6] – we consider the latter approach. For lack of a better term, we call it *worst-case redundancy.*

In [1] a *structural* characterization of all maximal solid codes contained in $a^+b^+ \cup a^+b^+a^+b^+$ is provided. With some effort it may be possible to extend this characterization to maximal solid codes in $(a^+b^+)^+$. The structural characterization permits us to determine lower bounds on the worst-case redundancy of maximal solid codes in $a^+b^+ \cup a^+b^+a^+b^+$.

An *algorithmic* characterization of all maximal finite solid codes is given in [4]. At this point we do not know how the algorithmic construction can be translated into the computation of the redundancy of the codes thus constructed.

Our paper is structured as follows: in Section 2 we briefly review the notions and notation used. Section 3 re-states the characterization, proved in [1], of the maximal solid codes in $a^+b^+ \cup a^+b^+a^+b^+$ in a fashion which is both more concise and more manageable. The crucial new notion is that of pairs of *near-inverses* of integer functions. The maximal solid codes in $a^+b^+ \cup a^+b^+a^+b^+$ are uniquely determined by such pairs of functions. Consequently, in Section 4, we derive some properties of near-inverses; moreover, for deriving bounds on the redundancy of solid codes we need continuous and strictly monotonically increasing pairs of such functions. In Section 5, we first show that the achievable redundancy of maximal solid codes in $a^+b^+ \cup a^+b^+a^+b^+$ is bounded from below by $\sqrt{2n} - \log n$ and exhibit an example that has this redundancy. We then provide a general technique for determining the redundancy of maximal solid codes in $a^+b^+ \cup a^+b^+a^+b^+$ from the pairs of near-inverse functions defining them. Section 6 contains a few concluding observations.

2 Notions and Notation

Let $\mathbb{N}$, $\mathbb{R}$ and $\mathbb{R}^+$ denote the sets of positive integers, of real numbers and of positive real numbers, respectively. We write $\mathbb{N}_0$ for $\mathbb{N} \cup \{0\}$, $\mathbb{N}_\infty$ for $\mathbb{N} \cup \{\infty\}$ and $\mathbb{R}^+_\infty$ for $\mathbb{R}^+ \cup \{\infty\}$. Open and closed intervals of real numbers are written as usual; for example $(x, x']$ denotes the left-open, right-closed interval $\{r \mid x < r \leq x'\}$.

Consider functions f and g of $\mathbb{R}^+$ into $\mathbb{R}^+$. We say that

$$f(x) \lesssim g(x) \ \textit{as}\ x \to \infty$$

if $\limsup_{x\to\infty} f(x)/g(x) \leq 1$; similarly,

$$f(x) \sim g(x) \ \textit{as}\ x \to \infty$$

if $\lim_{x\to\infty} f(x)/g(x) = 1$. We write $f \leq g$ to denote the fact that $f(x) \leq g(x)$ for all $x \in \mathbb{R}^+$. The function f is said to be *sub-linear* if $x \neq O(f(x))$; it is *supra-linear* if $f(x) \neq O(x)$.

An alphabet is a finite nonempty set. To avoid trivial cases, we assume that every alphabet has at least two elements, a and b. Let X be an alphabet. Then X^* is the set of *words* over X including the *empty word* ε and $X^+ = X^* \setminus \{\varepsilon\}$. The length of a word $w \in X^*$ is denoted by $|w|$ and, for $k \in \mathbb{N}_0$, X^k is the set of words of length k. A *language* over X is a subset of X^*. For a language L over X and for $k \in \mathbb{N}_0$, let $n_L(k)$ be the number of words of length k in L and let $N_L(k) = \sum_{i=0}^{k} n_L(i)$.

Let $L \subseteq X^+$ be an infinite language over X. An *L-encoding* is a bijection $\kappa : \mathbb{N}_0 \to L$ such that, for all n, $|\kappa(n)| \leq |\kappa(n+1)|$. The *worst-case redundancy of* κ (or of L) is the function

$$\varrho_\kappa(n) = \begin{cases} |\kappa(n)| - \lfloor \log n \rfloor - 1, & \text{if } n \geq 1, \\ |\kappa(n)| - 1, & \text{if } n = 0, \end{cases}$$

where the logarithm is taken with respect to base $|X|$. In the sequel, by *redundancy* we mean worst-case redundancy.

The asymptotic behaviour of the redundancy is determined by the values of n at which $|\kappa(n)|$ changes and by the growth of $|\kappa(n)|$ at these steps.

3 Maximal Solid Codes in $a^+b^+ \cup a^+b^+a^+b^+$

We review the characterization of all maximal solid codes in $a^+b^+ \cup a^+b^+a^+b^+$. The presentation of the characterization in this section is more concise than that of [1] and emphasizes the fact that these maximal solid codes are in a one-to-one correspondence with certain monotonically increasing partial functions of $\mathbb{N}$ into $\mathbb{N}$. First, we set up some required notation and prove some auxiliary results.

For $m \in \mathbb{N}_\infty$, let $I_m = \{i \mid i \in \mathbb{N}, i < m\}$. Thus, $I_1 = \emptyset$. As usual, let $\infty + 1 = \infty$. Consider a monotonically increasing function[1] $f : I_m \to \mathbb{N}$. When $m = 1$ then f is the empty function. Define $\lim f$ as

$$\lim f = \begin{cases} 0, & \text{if } f \text{ is empty,} \\ \infty, & \text{if } f \text{ is unbounded,} \\ p, & \text{if } f \text{ is bounded by } p \in \mathbb{N} \text{ and } f(j) = p \text{ for some } j \in I_m. \end{cases}$$

Definition 3.1 Let $m, n \in \mathbb{N}_\infty$ and let $f : I_m \to \mathbb{N}$ and $g : I_n \to \mathbb{N}$ be monotonically increasing functions. Then f and g are said to be (m,n)-*near-inverses of each other* if the following conditions are satisfied:

1. If $n = \infty$, then $\lim g = m$.
2. If $m = \infty$, then $\lim f = n$.
3. For all $i \in I_n$,
$$g(i) = \begin{cases} \min\{j \mid j \in I_m, f(j) > i\}, & \text{if such a } j \text{ exists,} \\ m, & \text{otherwise.} \end{cases}$$
4. For all $j \in I_m$,
$$f(j) = \begin{cases} \min\{i \mid i \in I_n, g(i) > j\}, & \text{if such an } i \text{ exists,} \\ n, & \text{otherwise.} \end{cases}$$

Let F_i and G_j be the sets used in conditions (3) and (4) in Definition 3.1, that is,

$$F_i = \{j \mid j \in I_m, f(j) > i\} \text{ and } G_j = \{i \mid i \in I_n, g(i) > j\}.$$

[1] The function f need not be strictly monotonically increasing.

The conditions (3) and (4) could, at a first glance, lead to $g(i) = \infty$ or $f(j) = \infty$ in some cases, which is, of course, not allowable. We verify that this cannot happen. Suppose $m = \infty$. Then $\lim f = n$ by (2). If also $n = \infty$ then the set F_i is non-empty for all $i \in I_n$. If $n < \infty$ then $f(j_0) = n$ for some $j_0 \in I_m$ and, hence, for $i \in I_n$ one has $f(j_0) > i$ and this shows again that all sets F_i are non-empty. Thus, in both cases g is well-defined. Similarly, one proves that f is well-defined.

Proposition 3.1 *Let $m, n \in \mathbb{N}_\infty$.*

1. *If f is a monotonically increasing function, $f : I_m \to \mathbb{N}$, bounded by n and such that $\lim f = n$ when $m = \infty$, then there is a unique monotonically increasing function $g : I_n \to \mathbb{N}$ such that f and g are (m, n)-near-inverses of each other.*

2. *If g is a monotonically increasing function, $g : I_n \to \mathbb{N}$, bounded by m and such that $\lim g = m$ when $n = \infty$, then there is a unique monotonically increasing function $f : I_m \to \mathbb{N}$ such that f and g are (m, n)-near-inverses of each other.*

3. *If f and g are (m, n)-near-inverses of each other then g and f are (n, m)-near-inverses of each other.*

For $m, n \in \mathbb{N}_\infty$ and a monotonically increasing function $f : I_m \to \mathbb{N}$, bounded by n and such that $\lim f = n$ when $m = \infty$, let $g_{f,m,n}$ be the function determined by Proposition 3.1(1).

We now state the characterization of all maximal solid codes in $a^+b^+ \cup a^+b^+a^+b^+$ proved in [1] using our new more concise terminology.

Theorem 3.1 ([1]) *Let $C \subseteq a^+b^+ \cup a^+b^+a^+b^+$.*

1. *The language C is a finite maximal solid code if and only if there are $m, n \in \mathbb{N}$ and a monotonically increasing function $f : I_m \to I_{n+1}$ such that*
$$C = \{a^n b^m\} \cup \{a^{f(j)} b^j a^i b^{g_{f,m,n}(i)} \mid j \in I_m, i \in I_n\}.$$

2. *The language C is an infinite maximal solid code if and only if there are $m, n \in \mathbb{N}_\infty$, at most one of which is finite, and a monotonically increasing function $f : I_m \to I_{n+1}$ with $\lim f = n$ when $m = \infty$ such that*
$$C = \{a^{f(j)} b^j a^i b^{g_{f,m,n}(i)} \mid j \in I_m, i \in I_n\}.$$

Consider $m, n \in \mathbb{N}_\infty$ and $f : I_m \to \mathbb{N}$. We say that the condition $S(f, m, n)$ is satisfied if and only if f is monotonically increasing, bounded by n, and $\lim f = n$ when $m = \infty$. By Theorem 3.1, the maximal solid codes in $a^+b^+ \cup a^+b^+a^+b^+$ are in a one-to-one correspondence with the triples (f, m, n) satisfying $S(f, m, n)$. For any such triple, let $C_{f,m,n}$ be the corresponding maximal solid code.

4 Properties of Near-Inverses

We explore the connection between functions that are near-inverses of each other in greater detail. The results will help us to compute bounds on the redundancy of solid codes.

Near-inverses are, really, nearly inverse functions of each other. To make this connection explicit, we resort to the tools of real calculus and specifically to continuous real functions.

For $m \in \mathbb{N}_\infty$, let $\mathbb{I}_m$ be the open real interval $(0, m)$. Then $\mathbb{I}_m \cap \mathbb{N} = I_m$. For any real function $f : \mathbb{I}_m \to \mathbb{R}^+$, let $f^* : I_m \to \mathbb{N}$ be the function defined by $f^*(j) = \lceil f(j) \rceil$ for $j \in I_m$.

Lemma 4.1 *For $m \in \mathbb{N} \cup \{\infty\}$ let $f : \mathbb{I}_m \to \mathbb{R}^+$ be a function. Then f^* is unbounded if and only if f is unbounded on $(0, m-1]$. Moreover, f^* is monotonically increasing when f is monotonically increasing.*

The following simple observation provides a key to the manipulation of near-inverses.

Lemma 4.2 *Consider $m \in \mathbb{N}_\infty$ and a monotonically increasing function $f : I_m \to \mathbb{N}$. Then there is a continuous strictly monotonically increasing function $\hat{f} : \mathbb{I}_m \to \mathbb{R}^+$ with the following properties:*

1. $\hat{f}^* = f$.

2. *For all $n \in \mathbb{N}$ one has $f(j) < n$ for all $j \in I_m$ if and only if $\hat{f}(x) < n$ for all $x \in \mathbb{I}_m$.*

3. $\lim f = \lim_{x \to m} \hat{f}(x)$.

Let $m \in \mathbb{N}_\infty$ and let $f : I_m \to \mathbb{N}$ be a monotonically increasing function. Any continuous strictly monotonically increasing real function $\hat{f} : \mathbb{I}_m \to \mathbb{R}^+$ satisfying the conditions of Lemma 4.2, is a *continuous strictly monotonically increasing approximation* of f. The precise shape of $\hat{f}$ can be important for the details of some of the bounds to be derived below. However, for expressing properties of near-inverses any such approximation $\hat{f}$ will be sufficient. Hence, without further mention we use $\hat{f}$ to denote an arbitrary continuous strictly monotonically increasing approximation of f.

Using these tools we obtain continuous strictly monotonically increasing approximations even for arbitrary monotonically increasing functions of $\mathbb{I}_m$ into $\mathbb{R}^+$ in the following sense. For $\varphi : \mathbb{I}_m \to \mathbb{R}^+$, consider $f = \varphi^*$. Then $\hat{f}^* = f = \varphi^*$; hence we define $\hat{\varphi} = \widehat{\varphi^*}$.

Let $m, n \in \mathbb{N}_\infty$ and let f be a continuous strictly monotonically increasing function of $\mathbb{I}_m$ into $\mathbb{R}^+$, bounded by n, such that $\lim_{x\to m} f(x) = n$ when $m = \infty$. In this case we say that $S(f, m, n)$ is satisfied. We define $g_{f,m,n} = g_{f^*,m,n}$ and $C_{f,m,n} = C_{f^*,m,n}$.

Proposition 4.1 *Let $m, n \in \mathbb{N}_\infty$ and let $f : \mathbb{I}_m \to \mathbb{R}^+$ be a continuous strictly monotonically increasing function such that $n, m > 1$ and $S(f, m, n)$ is satisfied. Then*

$$g_{f,m,n}(i) = \begin{cases} 1, & \textit{if } i < f(1), \\ 1 + \lfloor f^{-1}(i) \rfloor, & \textit{if } f(1) \le i < \lim_{x\to m} f(x), \\ n, & \textit{if } i \ge \lim_{x\to m} f(x), \end{cases}$$

for all $i \in I_n$.

The case distinction in Proposition 4.1 is necessary as one cannot assume in general that $I_n \subseteq f(\mathbb{I}_m)$ as would be required for $f^{-1}(i)$ to exist for all $i \in I_n$. While stated in terms of real functions, Proposition 4.1 is intended to be used mainly for integer functions. For example, if $f(j) = aj^2$ for some $a \in \mathbb{N}$ then $f^* = f$ and

$$g_{f,m,n}(i) = \left\{ \begin{array}{ll} 1, & \text{for } i < a \\ 1 + \left\lfloor \sqrt{i/a} \right\rfloor, & \text{for } i \ge a \end{array} \right\} = 1 + \left\lfloor \sqrt{i/a} \right\rfloor .$$

Similarly, if $f(j) = 2^j$ then

$$g_{f,m,n}(i) = \left\{ \begin{array}{ll} 1, & \text{for } i = 1 \\ 1 + \lfloor \log_2 i \rfloor, & \text{for } i > 1 \end{array} \right\} = 1 + \lfloor \log_2 i \rfloor .$$

The main advantage in considering real functions is that we obtain an explicit connection between near-inverses. Moreover, near inverses are nearly inverses of each other in the usual sense.

Proposition 4.2 *Let $m, n \in \mathbb{N}_\infty$ and let $f, h : I_m \to \mathbb{N}$ be monotonically increasing functions such that $S(f, m, n)$ and $S(h, m, n)$ are satisfied. If $f \le h$ then $g_{f,m,n} \ge g_{h,m,n}$ and conversely.*

Corollary 4.1 *Let $m, n \in \mathbb{N}_\infty$ and let $f, h : \mathbb{I}_m \to \mathbb{R}^+$ be continuous strictly monotonically increasing functions such that $S(f, m, n)$ and $S(h, m, n)$ are satisfied. If $f \leq h$ then $g_{f,m,n} \geq g_{h,m,n}$.*

Theorem 4.1 *Let $m, n \in \mathbb{N}_\infty$ and let $f : I_m \to \mathbb{N}$ be a monotonically increasing function such that $S(f, m, n)$ is satisfied. Let $g = g_{f,m,n}$. Then either both f and g are bounded by strictly monotonically increasing positive linear real functions from above and below or one of them is supra-linear and the other one is sub-linear.*

Theorem 4.1 reflects the fact that the functions f and g defining a maximal solid code in $a^+b^+ \cup a^+b^+a^+b^+$ are essentially inverses of each other. Hence, if f grows faster than any linear function, then g has to grow more slowly than any linear function and vice versa. This will imply further below that the maximal solid codes in $a^+b^+ \cup a^+b^+a^+b^+$ with smallest redundancy are those for which both f and g are linearly bounded from above and below.

5 Redundancy of Maximal Solid Codes in $a^+b^+ \cup a^+b^+a^+b^+$

We now show that the redundancy of any maximal solid code in $a^+b^+ \cup a^+b^+a^+b^+$ is asymptotically bounded from below by $\sqrt{2n} - \log n$. Moreover, this bound is tight.

Theorem 5.1 *Let $f : \mathbb{N} \to \mathbb{N}$ be an unbounded monotonically increasing function such that $S(f, \infty, \infty)$ holds true, let $C = C_{f,\infty,\infty}$ be the infinite maximal solid code in $a^+b^+ \cup a^+b^+a^+b^+$ defined by f, and let κ be a C-encoding. Then*

$$\sqrt{2n} \lesssim |\kappa(n)|$$

as $n \to \infty$.

The lower bound obtained in Theorem 5.1 is tight as shown by the following example.

Example 5.1 Let f and g be the functions of $\mathbb{N}$ into $\mathbb{N}$ defined by

$$f(n) = 2\left\lceil \frac{n}{2} \right\rceil - 1 \text{ and } g(n) = 2\left\lceil \frac{n}{2} \right\rceil + 1$$

for all $n \in \mathbb{N}$. Then f and g are (∞, ∞)-near-inverses of each other.

Let κ be a $C_{f,\infty,\infty}$-encoding. One computes that

$$-\frac{3}{2}+\sqrt{2n+\frac{137}{4}} \leq |\kappa(n)| \leq \frac{5}{2}+\sqrt{2n+\frac{121}{4}}$$

for all $n \in \mathbb{N}_0$. Hence $\kappa(n) \sim \sqrt{2n}$ as $n \to \infty$.

The basic idea of the proof of Theorem 5.1 is as follows: one considers $l, n \in \mathbb{N}$ such that

$$N_C(l) \leq n < N_C(l+1).$$

Then $|\kappa(n)| = l + 1$ and, hence, $\varrho_\kappa(n) = l - \lfloor \log n \rfloor$. Thus, one can determine the redundancy by solving the inequality above for l. This idea can be exploited to yield a more general lower bound, parameterized by f, on the redundancy of C.

For a continuous function $\varphi : \mathbb{R} \to \mathbb{R}$ and a constant $c \in \mathbb{R}$, let $[I_c\varphi]$ be the real function defined by

$$[I_c\varphi](x) = x - c + \int_c^x \varphi(t - c + 1)dt$$

whenever the integral exists.

Theorem 5.2 *Let $f : \mathbb{N} \to \mathbb{N}$ be a monotonically increasing function such that $n = O(f(n))$ and $S(f, \infty, \infty)$ holds true, and let $\hat{f}$ be a continuous strictly monotonically increasing approximation of f. Let $C = C_{f,\infty,\infty} = C_{\hat{f},\infty,\infty}$ and let κ be a C-encoding. Then there is a constant $c \in \mathbb{R}^+$ such that*

$$\left[I_c\hat{f}^{-1}\right]^{-1}(n) \lesssim |\kappa(n)|$$

as $n \to \infty$.

The bound in Theorem 5.2 is essentially tight in the following sense: the growth rate of f determines the growth rate of κ. For example, if f grows as n^k then κ grows as $\Omega(\sqrt[k+1]{n^k})$; if f grows as 2^n then κ grows as $\Omega(n)$.

6 Concluding Remarks

Lower bounds on the growth rate of the redundancy of infinite maximal solid codes in $a^+b^+ \cup a^+b^+a^+b^+$ can be derived directly from the near-inverse functions f and g defining such codes. Our presentation is restricted to the case when both $\lim f = \infty$ and $\lim g = \infty$, but can be easily extended to the case of only one of the limits being infinite. We suspect that these lower bounds are nearly tight and hence would provide useful lower bounds on the achievable redundancy of arbitrary solid codes.

References

[1] H. Jürgensen, M. Katsura, S. Konstantinidis: Maximal solid codes. *Journal of Automata, Languages and Combinatorics* **6** (2001), 25–50.

[2] H. Jürgensen, S. Konstantinidis: Codes. In G. Rozenberg, A. Salomaa (editors): *Handbook of Formal Languages*, 1. 511–607. Springer-Verlag, Berlin, 1997.

[3] S. Konstantinidis: Error-detecting properties of languages. *Theoret. Comput. Sci.* (to appear).

[4] N. H. Lâm: Finite maximal solid codes. *Theoret. Comput. Sci.* **262** (2001), 333–347.

[5] V. I. Levenshtein: Decoding automata, invariant with respect to the initial state. *Problemy Kibernet.* **12** (1964), 125–136, in Russian.

[6] V. I. Levenshtein: Maximum number of words in codes without overlaps. *Problemy Peredachi Informatsii* **6**(4) (1970), 88–90, in Russian. English translation: *Problems Inform. Transmission* **6**(4) (1973), 355–357.

[7] H. J. Shyr, S. S. Yu: Solid codes and disjunctive domains. *Semigroup Forum* **41** (1990), 23–37.

Maximal Independent Sets in Certain Subword Orders

Nguyen Huong Lam
Hanoi Institute of Mathematics
P. O. Box 631, Bo Ho, 10 000 Hanoi , Vietnam
Email: nhlam@ thevinh.ac.vn

1 Introduction

The free monoid, or the set of words on an alphabet, can be made into a poset (patially ordered set) by different kind of orders, one example in view is the lexicographic arrangement of words in a dictionary. One can also order the words by their length: the shorter words precede the longer ones in this order. The structure of an odered set is usually determined by its chains in the one hand and antichains in the other hand. Antichain — a notion from the Theory of Posets — is, as the term suggests, a subset in which no two distinct elements form a chain. In this article the antichains are called independent sets and we are concerned with independents sets in some specific orders on words.

The systematic study of independent subsets in the binary relation was initiated by Thierrin and Shyr probably in the mid 1970s [11]. Relations in consideration are such that their independent sets are usually codes [10], a subject which is treated thoroughly in [1], and it turns out that many fundamental classes of codes are that of independent subsets of certain simple binary relations, for instance, prefix, sufix, bifix, infix codes, hypercodes, uniform codes and so on.

In this paper we consider chiefly the classes of infix codes and hypercodes. By nature they both are the independent set in a class of very common orders compatible with the word length. Let us be precise as follows. Let A be a finite alphabet comprising a finite number of letters and A^* be the set of words defined on A, which is a free monoid on the basis A with the concatenation of words as product and the empty word, denoted 1, as the unit. For a set S and a word u of A^* we denote by $|S|$ the number of elements, or the cardinality, of S and by $|u|$ the length, or the number of letters, of u. By convention, $|1| = 0$

and by definition $|uv| = |u| + |v|$ for all words u, v. The word u is said to be a *factor* of v if $v = xuy$ for some words x, y of A^*. In particular, u is a *prefix* of v if $x = 1$, *suffix* if $y = 1$.

For any abstract sets S, T we use the notations $S - T$ and $S + T$ to denote their difference and union respectively; and for any subsets of words $X, Y \subseteq A^*$, we denote $XY = \{xy : x \in X, y \in Y\}$; $X^n = XX...X$ (n times) for a possitive integer n and $X^* = 1 + X + X^2 + \cdots$ the star closure of X. For a singleton set $\{w\}$ we use w^n, w^* instead of $\{w\}^n, \{w\}^*$.

We define the notion of subword; u is a *subword* of v if u is obtained from v by omitting some letters occurring in v, possibly none; formally, if

$$v = v_1 u_1 v_2 u_2 ... v_k u_k v_{k+1}$$

and

$$u = u_1 u_2 ... u_k$$

for a nonnegative integer k and some words $v_1, v_2, ..., v_k, v_{k+1}, u_1, u_2, ..., u_k$ in A^*.

For a binary relation on A^*, a subset of words is said to be *independent* if every pair of elements of it is incomparable by the relation. Clearly, we have a standard statement that, by Zorn's lemma, every independent set is included in a *maximal* (by inclusion) independent one. We deal exclusively with orders on words in the sequel. Recall that order is an reflexive, antisymmetric and transitive binary relation.

Obviously, the binary relations "being a factor of, a prefix of, a suffix of" and "being a subword of " all define partial orders on A^*, which we call correspondingly the factor, prefix, suffix and subword orders.

Independent sets in the prefix and suffix orders are called prefix and suffix sets respectively; independent sets in the factor order is called infix codes and that in the subword order are usually called hypercodes ([11]). The bifix relation is defined by means of disjunction of the prefix and suffix relations, independent sets in which are bifix codes. Otherwise speaking, a subset is a bifix code if none of its words is prefix or suffix of the others. While the prefix and suffix relations are orders, the bifix relation is not, since it is not transitive.

It has been known that every finite infix code may be embedded into a maximal infix code which is finite (see [7]) and that every hypercode is always finite (see [6], or [10] for a proof). The main scope of this paper is to describe a convenient method to construct all finite maximal infix codes as well as all maximal hypercodes containing a given infix and hypercode, but the exposition is done for independent sets of a more general class of orders, including the factor and subword orders, and certainly, the prefix and suffix orders as well. We highlight some definitions.

A binary relation $\prec$ on a subset L of A^* is called *length-compatible* if it is length-monotonic, i.e. for all words $u, v \in L$, $u \prec v$ implies $u = v$ or $|u| < |v|$ and, moreover, for every integer i : $|u| < i < |v|$ there exists a word w of

L such that $u \prec w \prec v$ and $|w| = i$. It is at once clear that the prefix, suffix, factor and subword orders, defined on $L = A^*$, are all length-compatible. Length-monotonic relation is considered in Van [13] where he has shown how to complete every finite independent set in such relations to a finite maximal one. I would like to thank Do Long Van for drawing the approach taken upon here to my attention.

The presentation is as folows. In the Section 2 below, we describe a procedure to obtain all the finite independent sets containing a given one in a length-compatible order. In particular this allows to obtain all finite independent sets of A^*. In §3 and §4 we apply the procedure to the concrete cases of infix code and hypercode.

The main idea of the construction originates from [8] but the argumentation is purified and more concentrated that now makes the exposition transparent and more unified in style.

We should say that the techniques in this papers could not be applied to bifix codes, but there is no need to do so since there has been already a remarkable work of Césari [2] giving an algorithm to determine all finite maximal bifix codes.

2 Maximal independent sets in length-compatible order

Let $\prec$ be a length-compatible order on a subset L of A^* and u, v be words of L. We say that u is a *predecessor* of v, or v is a *successor* of u, if $u \neq v$ and $u \prec v$; u is a predecessor (successor) of a subset X of A^* if u is a predecessor (successor) of a word in X. We denote the set of predecessors (successors) of X by $\mathrm{Pred}(X)$ ($\mathrm{Suc}(X)$ resp.). We also use $\mathrm{Pred}(X, Y)$ and $\mathrm{Suc}(X, Y)$ to denote the collection of predecessors and successors of X in Y respectively.

A subset Y of L is said to be a *base* of L if Y contains no distinct comparable words and every word of L is either in Y or is a successor or predecessor of Y. Otherwise speaking, a base is a maximal independent set of L with respect to $\prec$. A base is called initial if it has no prodecessors, and final if it has no successors.

It is obvious that every subset L posseses a unique initial base which is

$$Y_0 = \{u \in L : \mathrm{Pred}(u) = \emptyset\},$$

the collection of elements in L without predecessors. We pay special attention to those L having finite bases. For this purpose, it is enough to check Y_0 for finiteness.

Proposition 2.1 *A set possesses finite bases if and only if its initial base is finite.*

Proof The "if" direction is vacuously true. Let now L have a finite base Y. Every word of Y_0 is either in Y or in Pred (Y) as Y_0 has no successors in L and Y is a base. Thus every word of Y_0 has length not greater than the maximum length of words of Y, hence Y_0 is finite as the base Y and the alphabet A both are finite.

Our ultimate goal in this section is to determine all finite maximal independent sets containing a given one, but as the first step we consider the question to a larger extent: describe an efficient method to obtain all finite bases (if exist) of an arbitary subset L from the initial base of it provided, indeed, the premises involved are given constructibly.

For a base Y of L, we call a subset W of Y *coherent* if all words of W are of the same length n and there is a word u of length $n+1$ of L such that the set of predecessors in Y of u is W and no non-empty proper subset of W has this property, or equivalently, $\emptyset \neq \text{Pred}(u, Y) \subseteq W$ implies $\text{Pred}(u, Y) = W$ for every word u of length $n+1$. Otherwise speaking, a non-empty subset W of Y is coherent if it is a minimal set with respect to the folowing : (i) all words of W have the same length n, and (ii) W has the form $\text{Pred}(u, Y)$ for a word u of length $n+1$. Certainly, not every base has coherent subsets but also not every base has no coherent subsets. To wit

Proposition 2.2 *A base has coherent subsets if and only if it is not final.*

Proof The direct implication is trivial. For the converse, we make an observation that if Y is a base, not final, of L then there exist $u \in L$ and $y \in Y$ such that $y \prec u$ which in turn implies that there exist $v \in L$ such that $y \prec v$ and $|v| = |y| + 1$ by length-compatibility of $\prec$. Let n be the smallest integer such that none of the words of L has predecessors in Y of length less than n but some do have predecessors of length n in Y. Actually, this integer exists by the observation above, hence the set of words of length $n+1$ having all predecessors in Y of length n is not empty. We choose among such words one with the minimal (by inclusion) set of predecessors in Y. It is easy to chek that these predecessors form a coherent subset of Y.

For a coherent subset W (of word-length n) of Y, denote $D(W, Y)$, or shortly $D(W)$ when there is no need of explicit reference to Y, the set of words of L of length $n+1$ all of whose predecessors in Y are precisely the words of W:

$$D(W, Y) = \{u \in L : \text{Pred}(u, Y) = W, |u| = n+1\}.$$

We define now the following transformation of Y:

$$F(Y, W) = Y - W + D(W, Y)$$

that will be the operation we need to produce a new base from a given one.

Proposition 2.3 *If Y is a (finite) base then $F(Y)$ is also a (finite* resp.*) base.*

Proof It is immediate to see that F preserves finiteness. We show that F preserves independence and maximality as well.

Independence: Two arbitrary distinct words of $D(W)$ are not comparable by $\prec$ as they are of equal length. The same is true for $Y - W$ as Y is independent. Every word of $D(W)$ has only predecessors in W and no successors in Y as it has predecessors in Y (in W, to be exact) already, which means that every word of $D(W)$ is not comparable to any word in $Y - W$. This shows that $F(Y, W) = Y - W + D(W)$ is independent.

Maximality: Suppose that $u \in L$ and $u \notin F(Y)$ (sometimes we omit the reference to W in $F(Y, W)$ for brevity). We show that u has either predecessors or successors in $F(Y)$. If $u \in Y$ then $u \notin W$ as $u \in Y - W \subseteq F(Y)$ which shows that u is a predecessor of $D(W)$, or just the same, u has successors in $D(W) \subseteq F(Y)$. If $u \notin Y$ we have two cases:

(i) u has predecessors in Y. If u, moreover, has predecessors in $Y - W \subseteq F(Y)$ then this alternative is done. If, otherwise, every predecessor of u in Y is in W then $|u| > n + 1$, therefore u has a predecessor u_0 with $|u_0| = n + 1$ such that all predecessors of u_0 in Y are in W, hence $\text{Pred}(u_0, Y) = W$ by coherency of W that means $u_0 \in D(W)$ that is u has a predecessor in $D(W) \subseteq F(Y)$.

(ii) u has successors in Y. Then u has successors in $Y - W$ or else in W, therefore, in $D(W)$. In either case u has successors in $F(Y)$. This completes the proof.

We now define another operation, reverse to F which is instrumental in proving that multiple use of F can engender all finite bases, but some notions first.

A predecessor of Y is said to be *direct* if it has no successors among the predecessors of Y or in other words, it is not a predecessor of any predecessors of Y. We call a system S of direct predecessors of Y *cohesive* if the words of S are of the same length n and the set of successors in Y of each of them is the same (which is actually $\text{Suc}(S, Y)$) and, moreover, for any predecessor u of $\text{Suc}(S, Y)$ if $|u| = n$ and $u \notin S$ then $\text{Suc}(u, Y) - \text{Suc}(S, Y) \neq \emptyset$. The following reformulation is more intuitive. A system S of direct predecessors is cohesive if and only if (i) it consists of the words having the same set of successors in Y, and (ii) $\text{Suc}(S, Y)$ is minimal among the subsets of Y which are the the set of successors of a set satisfying the preceding condition. The equivalence of this definition to the former one is essentially shown in the converse of the proposition below. Note that every word of $\text{Suc}(S, Y)$ has length $n + 1$.

Does every base possess a cohesive system of direct predecessors? The answer is just the opposite extreme to the case of coherent subsets.

Proposition 2.4 *A finite base has no cohesive system of predecessors if and only if the set of its predecessors is empty, that is, if and only if it is the initial base.*

Proof The initial base has no predecessors by definition hence has no whatever ones.

For the converse, let Y be a finite base with (finite) non-empty collection of predecessors and once finite it admits at least one direct predecessor. Consider a direct predecessor s_1 with the property that $\mathrm{Suc}(s_1, Y)$ is minimal by inclusion among its counterpart $\mathrm{Suc}(s, Y)$ when s run through the direct predecessors of Y. Denote S the set consisting of those predecessors s satifying $|s| = |s_1|$ and $\mathrm{Suc}(s, Y) \subseteq \mathrm{Suc}(s_1, Y)$. Certainly, S is not empty because it contains s_1. We show that S is cohesive.

In fact, for every $s \in S$ the inclusion $\mathrm{Suc}(s, Y) \subseteq \mathrm{Suc}(s_1, Y)$ implies that the predecessor s is direct. Moreover the minimality of $\mathrm{Suc}(s_1, Y)$ forces the equality $\mathrm{Suc}(s, Y) = \mathrm{Suc}(s_1, Y)$. For a predecessor s_2 of $\mathrm{Suc}(s_1, Y)$ with $|s_2| = |s_1|$ the fact that $s_2 \notin S$ means by definition that $\mathrm{Suc}(s_2, Y)$ is not a subset of $\mathrm{Suc}(s_1, Y)$, or equivalently, $\mathrm{Suc}(s_2, Y) - \mathrm{Suc}(s_1, Y) \neq \emptyset$. The cohesivity of S is proved.

Now we are ready to define the anticipated inverse of F: Let Y be a base of L and S a cohesive system of direct predecessors of Y. We define the transformation B as

$$B(Y, S) = Y - \mathrm{Suc}(S, Y) + S.$$

Of course we expect that B has the same property as F that is the content of the following

Proposition 2.5 *The transformation B preserves finiteness, independence and maximality of the base Y.*

Proof If Y is finite then $B(Y, S)$ (onward we use $B(Y)$ for short, when possible) is obviously finite, since S is finite. For the independence of $B(Y)$, we observe first that every pair of elements of S are not comparable since they are of the same length. Second, every pair of different words of $Y - \mathrm{Suc}(S, Y)$ are not comparable either as they are in the independent Y. Finally, every word of S has no predecessors in $Y - \mathrm{Suc}(S, Y)$ since it is a predecessor of $\mathrm{Suc}(S, Y) \subseteq Y$ and the order is transitive; further it has no successors in $Y - \mathrm{Suc}(S, Y)$ either since they all are already in $\mathrm{Suc}(S, Y)$ by definition.

Now we prove the maximality of $B(Y)$. Consider an arbitrary word u not belonging to $B(Y)$. If $u \in Y$ then $u \in \mathrm{Suc}(S, Y)$ then u has successor in $S \subseteq B(Y)$. Alternatively, suppose that u has a successors v in Y. The case $v \in \mathrm{Suc}(S, Y)$ ensures that v, hence u, has successor in S; the other case $v \in Y - \mathrm{Suc}(S, Y)$ shows that v itself belong to $B(Y)$.

Further, suppose that u has a predecessor w in Y. If $w \in Y - \mathrm{Suc}(S, Y)$ then $w \in B(Y)$; if $w \in \mathrm{Suc}(S, Y)$ then w in turn has successors in S, that is, again in $B(Y)$. Thus, in all instances we have shown that u has either successors or predecessors in $B(Y)$: $B(Y)$ is a base.

The following assertion shows that B is a right inverse of F. As a matter of fact, we could prove that B is really an inverse to F (both left and right) but we do not need that much in the sequel.

Proposition 2.6 *Let Y be a base with S a cohesive system of its direct predecessors. Then S is a coherent subset of $B(Y,S)$ with $D(S,B(Y,S)) = \mathrm{Suc}(S,Y)$ and $Y = F(B(Y,S),S)$.*

Proof Put $Y' = B(Y,S) = Y - \mathrm{Suc}(S,Y) + S$. Of course $S \subseteq Y'$; we shall show it to be a coherent subset of Y'. By definition, all words of S have the same length n; we first show that there exists a word of length $n+1$ the set of predecessors of which in Y' is precisely S. Take an arbitrary word u of $\mathrm{Suc}(S,Y)$ (certainly $|u| = n+1$). It follows that $S \subseteq \mathrm{Pred}(u)$ as S is a cohesive system for Y and $u \in Y$, hence $S \subseteq \mathrm{Pred}(u,Y')$ as $S \subseteq Y'$. Next, if $v \in \mathrm{Pred}(u,Y')$ then $v \notin Y$ as $u \in Y$, hence $v \in Y' - Y \subseteq S$. That is $\mathrm{Pred}(u,Y') \subseteq S$ and the equality $\mathrm{Pred}(u,Y') = S$ follows for every $u \in \mathrm{Suc}(S,Y)$.

Further, we assume that $\emptyset \neq \mathrm{Pred}(u,Y') \subseteq S$ for a word u of length $n+1$. If u admits a predecessor v in Y, we have $|v| < |u| = n+1$, or $|v| \leq n$, which implies that v cannot be a successor of S, so $v \in Y - \mathrm{Suc}(S,Y) \subseteq Y'$. Hence $v \in \mathrm{Pred}(u,Y') \subseteq S$ that contradicts the fact that $v \in Y$ and the words of S are predecessors of Y. Consequently, u has no successors in Y. But u has no sucessors in Y either since u has a predeccessor in S and this predeccessor is a direct one of Y. As Y is a base, it remains for u to be in Y, hence $u \in \mathrm{Suc}(S,Y)$. Now that u is a successor in Y of some word of S, and S is a cohesive system for Y then u is a successor in Y of all words of S, or equivalently, every word of S is predecessor of u in Y', that is $S \subseteq \mathrm{Pred}(u,Y')$. Combining with the assumption we get $\mathrm{Pred}(u,Y') = S$.

Stated briefly, we have proved that if $|u| = n+1$ and $\mathrm{Pred}(u,Y') \neq \emptyset$ then $\mathrm{Pred}(u,Y') \subseteq S$ if and only if $\mathrm{Pred}(u,Y') = S$ if and only if $u \in \mathrm{Suc}(S,Y)$. This shows that S is a coherent subset of Y' and, moreover, $D(S,Y') = \mathrm{Suc}(S,Y)$. Eventually we have

$$F(B(Y,S),S) = F(Y',S) = Y' + D(S,Y')$$

$$= Y - \mathrm{Suc}(S,Y) + S - S + D(S,Y') = Y.$$

as was to be proved. Note that in this proof we do not require the cohesivity at its full strength.

Now we describe a method to obtain all finite bases of L. We apply the transformation F to Y_0, with an appropriate coherent subset to obtain a new base Y_1, then again to Y_1 we apply F whenever possible to get Y_2, etc. That is we apply successively F to Y_0 with properly choosen coherent subsets of bases in process, in all possible ways, to obtain new and new bases. We state that in this manner we can obtain every finite base of L.

Theorem 2.7 *For every finite base Y of L there exists a sequence of finite bases $Y_0, Y_1, ..., Y_n$ with their corresponding coherent subsets $S_0, S_1, ..., S_n$, beginning with the initial base Y_0 and satisfying $Y_1 = F(Y_0, S_0)$, ..., $Y = F(Y_n, S_n)$.*

Proof Since Y is finite, the number of predeccessors of Y is also finite; the proof is proceeded by induction on the cardinality of $\text{Pred}(Y)$. If $\text{Pred}(Y) = \emptyset$ then Y is the initial base and the sequence we needed is empty. Let Y be a base and assume that the theorem is valid for all bases with fewer than $|\text{Pred}(Y)|$ predecessors. By Proposition 2.4, we dispose at least one cohesive system of direct predecessors of Y, say S. Then

$$Y' = B(Y, S) = Y - \text{Suc}(S, Y) + S$$

is a finite base with $\text{Pred}(Y') = \text{Pred}(Y) - S$. Consequently, the cardinality of $\text{Pred}(Y')$ being less than $\text{Pred}(Y)$, the induction hypothesis applied to Y' yields a sequense of bases $Y_0, Y_1, ..., Y_{n-1}$ and the corresponding coherent subsets $S_0, S_1, ..., S_{n-1}$ satifying

$$Y_1 = F(Y_0, S_0), ..., Y' = F(Y_{n-1}, S_{n-1}).$$

In virtue of Proposition 2.6, $Y = F(Y', S)$. Putting $Y_n = Y'$ and $S_n = S$, we get the sequence

$$Y_1 = F(Y_0, S_0), ..., Y_{n-1} = F(Y_{n-1}, S_{n-1}), Y = F(Y_n, S_n)$$

with the corresponding sequence of coherent subsets $S_0, ..., S_{n-1}, S_n$ as desired to prove.

It is noteworthy that the backward sequence need not be unique, thus a base can be obtained several times by the procedure.

Let now $\prec$ be a length-compatible order defined on the whole A^*. We now return to the task set up at the onset of this section: determine all the finite maximal independent sets containing an independent set X that turns out now to be a simplified question. Set

$$L(X) = A^* - \text{Pred}(X) - X - \text{Suc}(X)$$

as the set of the words not comparable with any word in X. We say that a set Y is a complement to X if Y is disjoint from X and $X + Y$ is a maximal independent set of A^*. In this case we say also that $X + Y$ is a completion of X. For a maximal independent set Z containing X, clearly $Z - X$ is a complement to X; our problem is equivalent to determining all complements to X.

It is now important that the order on $L(X)$ inherited from the order $\prec$ on A^* is also length-compatible. To see that it suffices to verify that for all words u, v, w of A^* the relation $u \prec w \prec v$ implies $w \in L(X)$ whenever $u, v \in L(X)$. In fact, if this is not the case, $w \notin L(X)$ means that w is either in X or is a successor or predecessor of X, hence either v is a successor of X or u is a predecessor of X, that is, either u or v is not in $L(X)$: a contradiction.

The following assertion shows that complements are nothing else but bases.

Proposition 2.7 *Y is a complement to X if and only if it is a base of $L(X)$.*

Proof It is trivial to see that if Y is a complement to X then $Y \subseteq L(X)$. Since $X+Y$ is a maximal independent set, every word of $L(X)$ must be in $X+Y$ or $\mathrm{Pred}(X+Y)$ or $\mathrm{Suc}(X+Y)$ but it is forbidden to be in $\mathrm{Pred}(X)+X+\mathrm{Suc}(X)$ so it must be in $\mathrm{Pred}(Y)+Y+\mathrm{Suc}(Y)$. This means that Y is a base of $L(X)$, as Y is independent. Conversely, let Y be a base of $L(X)$. Then, first, $X+Y$ is independent, Y is disjoint from X and, for an arbitrary word of A^*, if $u \in L(X)$ then $u \in \mathrm{Pred}(Y)+Y+\mathrm{Suc}(Y)$, as Y is a base; if $u \notin L(X)$ then $u \in \mathrm{Pred}(X)+X+\mathrm{Suc}(X)$. In either case $u \in \mathrm{Pred}(X+Y)+(X+Y)+\mathrm{Suc}(X+Y)$ showing that $X+Y$ is maximal independent.

Now let X futher be finite. Determination of the finite complements to X is just then the determination of all finite bases of $L(X)$. Of course, we cannot be certain that $L(X)$ always possesses finite bases, however, when the underlying order satisfies a mild condition, $L(X)$ does so.

Proposition 2.8 *If the length-compatible order $\prec$ is such that every non-empty word is a successor of the empty word then $L(X)$ possesses finite bases for every finite independent set X.*

Proof It suffices to show that the initial base Y_0 is finite. To do this we just bound above the length of words of Y_0.

Let $u \in Y_0$ and $u \neq 1$. Since $1 \prec u$, there exists a word $v \in A^*$ such that $v \prec u$ and $|v| = |u| - 1$. By definition, u has no predecessors in $L(X)$, hence $v \notin L(X)$, that is $v \in \mathrm{Pred}(X)+X+\mathrm{Suc}(X)$, but $v \notin X+\mathrm{Suc}(X)$ otherwise $u \in \mathrm{Suc}(v) \subseteq \mathrm{Suc}(X)$ in spite of $u \in L(X)$. Consequently $v \in \mathrm{Pred}(X)$ and $|v|$ is less than the maximal length of the words of X, which is finite as X is finite, and $|u|$ is less than that maximal value plus 1. The proposition is proved.

Our general consideration is finished, we go on to concrete examples in the two following sections. Although the prefix, suffix, factor and subword orders are all among the length-compatible ones, we treat only the cases of factor and subword order in detail, as the construction prefix and suffix codes are simple (by trees) and have been treated comprehensively elsewhere (cf. [1], [10]).

3 Factor order. Infix codes

We make a transcript of all done in the previous section for the factor order. Now a word is a predecessor of another if the first is a factor of the latter.

Independent sets in the factor order are called infix sets or infix codes. Infix sets happen to be finite or infinite.

Example 3.1 Consider a binary alphabet $A = \{a, b\}$. The following sets are maximal infix sets:
(a) $\{aa, aba, bab, bb\}$;
(b) $a^2b^2 + b^2a^2 + \{ba^ib : i = 1, 2, ...\} + \{ab^ia : i = 1, 2, ...\}$;
(c) $ba^2 + bab + \{a^ib^ia : i = 1, 2, ...\}$.

Certainly the factor order is length-compatible and, moreover, it satisfies Proposition 2.8 as the empty word is a factor of every word, thus $L(X)$ has finite bases for all finite infix sets X. In particular, this shows that every finite infix code has a finite completion that has been established in [7].

Let X be an infix set. It is straightforward to see that

$$L(X) = A^* - A^*XA^* - F(X)$$

where $F(X)$ denotes the set of factors of X, and the initial infix base of $L(X)$ is $I_0 = L(X) - A^*L(X)A^+ - A^+L(X)A^*$.

We now clarify what sets can be coherent subsets or cohesive systems for an infix bases I. Note that a word of length $n+1$ has at most two factors of length n which are the longest prefix and suffix of its, consequently, every coherent subset has one or two words.

If a coherent subset W of I is a singleton, $W = \{u\}$, then clearly the set of words of length $|u| + 1$ having just u as factors in I is

$$D(u, I) = L(X) \cap (((uA - A^*(I - u)) + (Au - (I - u)A^*)).$$

If W has two elements, $W = \{u, v\}$, then the corresponding $D(W, I)$ is

$$D(\{u, v\}, I) = L(X) \cap ((uA \cap Av) + (Au \cap vA))$$

and, of cousre, with $D(u, I) = \emptyset$ and $D(v, I) = \emptyset$ saying that $\{u\}$ and $\{v\}$ both are not coherent. Observe also that in this case $D(W, I)$ has two words at the most because $uA \cap Av$ is either empty or singleton. Summing up, we have

Proposition 3.2 *A subset W of the infix base I is coherent if and only if $W = \{u\}$ with non-empty $D(u, I) = L(X) \cap (((uA - A^*(I - u)) + (Au - (I - u)A^*))$ or $W = \{u, v\}$ with non-empty $D(\{u, v\}, I) = L(X) \cap ((uA \cap Av) + (Au \cap vA))$ but empty $D(u, I)$ and $D(v, I)$.*

Example 3.3 Let $A = \{a, b\}, X = \{a^2, b^2\}$. Then $L(X) = \{ab, ba\} + \{(ab)^i : i > 0\} + \{b(ab)^i : i > 0\} + \{a(ba)^i : i > 0\} + \{(ba)^i : i > 0\}$. The intial infix base $L(X)$ is $I_0 = \{ab, ba\}$. Calculation shows that $D(\{ab\}, I_0)$ and $D(\{ba\}, I_0)$ are both empty, so I_0 has no coherent singleton. We conlude at once that $W = I_0$ must be coherent, but we calculate however: $D(I_0, I_0) = L(X) \cap ((abA \cap Aba) + (Aab \cap baA)) = (aba + bab) \cap L(X) = aba + bab$.

Next, we investigate under which conditions a finite infix base is final. Remind that a finite infix base is final if and only if every word of $L(X)$ is a factor of the base. It follows that $L(X)$ is then finite too. The sets X with $L(X)$ finite are nothing but *unavoidable sets,* a subject in Combinatorics on Words investigated in detail in [3] or in a recent paper [9]. A subset of A^* is unavoidable if all but finitely many words of A^* have a factor in it. Of course not every unavoidable set is infix but the *minimal* unavoidable set (i.e. one if deprived of any one element is no more unavoidable) is infix.

The following assertion is true not only for the factor order but also for all length-compatible orders.

Proposition 3.4 *$L(X)$ has finite final bases if and only if $L(X)$ is finite.*

Proof The "only if" part is proved above. Let $L(X)$ be finite then starting from I_0 we cannot apply the transformation infinitely, that is, at some step, the base in consideration has no coherent subsets. This base is final.

Example 3.5 Let $A = \{a, b\}$ and $X = \{aa, bb, bab\}$. Then $L(X) = \{aba\}$ and trivially $L(X)$ has a unique infix base $I_0 = \{aba\}$ which is initial and final at the same time and which indicates also that the only finite maximal infix code containing $\{aa, bab, bb\}$ is $\{aa, bab, bb; aba\}$.

The Proposition 3.4 says also that when X is unavoidable there are only finitely many infix bases in $L(X)$, or equivalently, there are only finitely many infix codes containing X and vice versa. We further give an example illustrating the execution of the procedure.

Example 3.6 Consider the set in the Example 3.3, $X = \{a^2, b^2\}$. We have shown that the initial base of $L(X)$ is $I_0 = \{ab, ba\}$ and the whole I_0 is a (unique) coherent subset with $D(I_0, I_0) = \{aba, bab\}$. Thus $I_1 = F(I_0, I_0) = I_0 - I_0 + D(I_0, I_0) = \{aba, bab\}$ is an infix base of $L(X)$ and $\{a^2, b^2, aba, bab\}$ only is a maximal infix code containing $\{a^2, b^2\}$.

Concluding we should say that the important case $X = \emptyset$ with $L(X) = A^*$, when we have to determine all finite maximal infix codes of A^* has been treated in detail in [8].

4 Subword order. Hypercodes

Recall that u is a subword of v if

$$u = u_1 u_2 ... u_n v = v_1 u_1 ... v_n u_n v_{n+1}$$

for some (possibly empty) words $u_1, u_2, ..., u_n, v_1, v_2, ... v_n, v_{n+1}$ on A. It is also evident that the relation "u is a subword of v" is a length-compatible order. Hypercode is a set independent in this order. It is interesting to mention that while infix bases may be finite or infinite the hypercodes are definitely finite that is a theorem of Higman [6], one can see also [10] for a proof.

To define formally the set of words having subwords in a given set we can use the notion of *shuffle product* [5]. Let X and Y be sets of words, we denote by $X \circ Y$ the set of words of the form $x_1 y_1 x_2 y_2 ... x_n y_n x_{n+1}$ with $n \geq 0$ and $x_1 x_2 ... x_n x_{n+1} \in X$, $y_1 y_2 ... y_n \in Y$ for (possible empty) words $x_1, x_2, ... x_n, x_{n+1}, y_1, y_2, ... y_n$ of A^*. Now it is easy to see that the set of words having subwords in X is $X \circ A^*$. Let now X be a hypercode then $L(X) = A^* - X \circ A^* - S(W)$, where $S(W)$ stands for the set of subwords of W and the initial subword base of $L(X)$ is $Y_0 = L(X) - L(X) \circ A^*$. For a coherent subset S of a base Y we have clearly $D(S, Y) = L(X) \bigcap_{s \in S} s \circ A$. We consider an example to show the execution of procedure.

Example 4.1 Let $X = \{a^2, aba\}$ on the alphabet $\{a, b\}$. Then $L(X) = \{b^i : i > 1\} + \{b^i a b^j : i + j > 1\}$ and $Y_0 = \{b^2, bab\}$. The set $S_0 = \{b^2\}$ is coherent for Y_0 and $D(S_0, Y_0) = \{b^3, b^2 a, ab^2\}$. We get then a new base:

$$Y_1 = F(Y_0, S_0) = \{b^2, bab\} - \{b^2\} + \{b^3, b^2 a, ab^2\} = bab + b^3 + b^2 a + ab^2$$

of $L(X)$. Note that $\{bab\}$ is not a coherent set of Y_0 since all words of length 4 of $L(X)$ have equally b^2 as a subword.

We now state some specific properties of the subword order.

Proposition 4.2 *If a coherent subset S has at least three words, the corresponding set $D(S)$ has at most one word.*

The Proposition is an immediate consequence of the following lemma.

Lemma 4.3 Two distinct words of length $n + 1$ have at most two subwords of length n in common.

Proof Suppose on the contrary that two words w_1 and w_2 of length $n + 1$ have three common subwords. We can assume further that w_2 has two dictinct subwords $u_1 v_1$ and $u_2 v_2$ for which

$$w_2 = u_1 c v_1, \qquad w_2 = u_2 d v_2 \tag{1}$$

with $c, d \in A, u_1, v_1, u_2, v_2 \in A^*$ and

$$w_1 = u_1 v_{11} a v_{12}, \qquad w_1 = u_2 v_{21} b v_{22} \tag{2}$$

with $a, b \in A, u_1, v_{11}, v_{12}, u_2, v_{21}, v_{22} \in A^*$ and $v_1 = v_{11}v_{12}$, $v_2 = v_{21}v_{22}$.

Note that $|u_1| \neq |u_2|$, otherwise the two subwords are equal. Let suppose that $|u_2| < |u_1|$, hence $u_1 = u_2 t$ for $t \neq 1$. Then (1) and (2) give

$$tcv_1 = dv_2 = dv_{21}v_{22} \tag{3}$$

and

$$tv_{11}av_{12} = v_{21}bv_{22}. \tag{4}$$

The last equality implies that t or v_{21} is a prefix of the other. If t is a prefix of v_{21} then by (3) we get $tc = dt$, consequently we get $c = d$ and $t = d^k = c^k$ for $k > 0$. It follows that $w_2 = u_1 c v_1 = u_2 t d v_1 = u_2 d^{k+1} v_1 = u_2 d t v_1 = u_2 d v_2$ and then $v_2 = tv_1$ which yields, together with $u_1 = u_2 t$, that $u_1 v_1 = u_2 v_2$: a contradiction! If, otherwise, v_{21} is a proper prefix of t, again from (3) we get that v_{21} is a prefix of dv_{21}. Consequently, $v_{21} = d^k$, $k \geq 0$, hence d^{k+1} is prefix of t. But by (4) $d^k b$ is also a prefix of t, which forces $d = b$. Now this implies $w_1 = u_2 v_{21} b v_{22} = u_2 d^k b v_{22} = u_2 d^{k+1} v_{22} = u_2 d d^k v_{22} = u_2 d v_{21} v_{22} = u_2 d v_2 = w_2$: again contradiction. The proof is complete.

Unlike infix bases, we cannot say that all coherent subsets of subwords bases are one- or two-element sets. The following argument help constructing large coherent subsets.

We say that a word w has complexity k provided k is the smallest integer satisfying $w \in a_1^* \dots a_k^*$ for letters $a_1, ..., a_k \in A$. It is straightforward to verify that if w has complexity k then the subwords of length $|w|-1$ of w are exactly k in number and, besides, they all have complexity at least $k-2$. It is also evident that a word containing words of complexity k as subwords have complexity at least k. Let now w be a word with high complexity k (k is large) and of length $n+1$. Let denote the sets of subwords of length $n-1$ and of length n of w by U and S, recpectively. When w is "complex" the cardinality of S is certainly large. It is not complicated to verify that for a subword u of length $n-1$ of w there always exist a word of length n having u as subword and not being subword of w; we spare one of such words, denoted $s(u)$, by each $u \in U$. Now let Z be any subset of S comprising sufficiently many words and X the set of the remaining words of length n of A^*. Note that $s(u) \in X$ for all $u \in U$.

We see that, first, X is a hypercode (of words of the same length) and, second, Z is included in $L(X)$ and moreover, in the initial subword base Y_0 of $L(X)$ since every subword u of length $n-1$ of Z, being in U, is a subword of $s(u)$, hence of X. That is $u \notin L(X)$ and every word of Z has no subwords in $L(X)$, therefore $Z \subseteq Y_0$.

We show that Z is a coherent subset of Y_0. In fact, by construction Z is a subset of subwords of length n of w in Y_0, so the word of Z has complexity not less than $k-2$. Since a word of length n, if not in Z is in X, every word v of $L(X)$ has all subwords of length n in Z, hence it has complexity at least $k-2$ and the subwords of length n of v amount up to at least $k-2$ in number and they are all in Z. Consequently, when $|v| = n+1$, v has at least $k-2$ subwords of length n in common with w which implies that $v = w$ by Proposition 4.2 as k is large by choice. That is to say there is none but one word, namely w, of length $n+1$ whose subwords of length n are all in Z that shows the coherency of Z.

Thus every word of complextity 5 will be appropriate to produce large Z, but we shall see that sometimes $k = 3$ is also fit for our purposes. The following example shows the detail.

Example 4.4 Let $A = \{a, b\}$. Let take $w = ababa$ with $n+1 = 5$ and $k = 5$ then $S = \{abab, aba^2, ab^2a, a^2ba, baba\}$ is of four elements. Any subset Z of S containing more than two words is a desired coherent set for the initial subword base Y_0 of $L(X)$.

However, some words of lower complexity also bring large coherent sets S even for $k = 3$, the least admissible value of complexity for words in $D(S)$. Consider $w = a^2b^2a$, $n+1 = 5$, $k = 3$. Here is the set $S = \{ab^2a, a^2ba, a^2b^2\}$ of three elements. The only possible issue for Z to take is $Z = S$. It is straightforward to verify that all words of length 5, and of complexity 2, contain subwords out of Z i.e., in X, hence they are not in $L(X)$. Next, if their complexity, otherwise, are not 2 then they have subwords in Z only if their complexity is not 1, hence is at least 3. Direct inspection yields w as the only candidate that has no subwords in X. This shows that Z is coherent for the corresponding Y_0 of $L(X)$.

Now we turn to the question when $L(X)$ admits a final subword base? That is when is $L(X)$ finite? Again we can think of a kind of unavoidable set in the subword order. If $L(X)$ is finite the sufficiently long words are not in $L(X)$ that means they have subwords in X. Conversely, if X is unavoidable then the words in $L(X)$ avoid X therefore they are finite in number.

It appears that in the subword order the unavoidable set has a very simple characterization. It is clear that an unavoidable set contains a^n for each $a \in A$ and some $n > 0$. Conversely, if a set satisfies this property then the words avoiding it have the occurences of the letter a fewer than the corresponding integer n. Hence their length is bounded above by the sum of all these exponents n's and the sufficiently long words have subwords in the set. That is, a set is unavoidable in the subword order if and only if it contains a power of every letter. Summing up, we state an analogous result to the factor order.

Proposition 4.5 *Let X be a hypercode. Then $L(X)$ is finite if and only if one of the following equivalent conditions holds: X is unavoidable, X contains a power of every letter and there are finitely many maximal hypercodes containing X.*

For the special case when X is empty, $L(X)$ is the whole A^*, the initial subword base is A, our procedure allows to dertermine all maximal hypercodes of A^*. We conlude the paper with an example showing the performance in some detail.

Example 4.6 Let $A = \{a, b\}$ and the hypercode $X = \{a^2, ab, ba, b^3\}$ is maximal which is verified directly. X can be derived from A by the following sequence of applications of F:

$X_1 = \{a, b^2\} = F(A, S_0), S_0 = \{b\}, D(S_0, A) = \{b^2\}$,

$X_2 = \{a, b^3\} = F(X_1, S_1), S_1 = \{b^2\}, D(S_1, X_1) = \{b^3\}$,

$X = X_3 = \{ab, ba, a^2, b^3\} = F(X_2, S_2),\ S_2 = \{a\}, D(S_2, X_2) = \{ab, ba,\ a^2\}$.

This sequence, however, is not unique, X can be obtained by another sequence as follows:

$X_1 = \{a^2, b\} = F(A, S_0), S_0 = \{a\}, D(S_0, A) = \{a^2\}$,

$X_2 = \{a^2, ab, ba, b^2\} = F(X_1, S_1), S_1 = \{b\}, D(S_1, X_1) = \{ab, b^2, ba\}$

and

$X = X_3 = \{a^2, ab, ba, b^3\} = F(X_2, S_2), S_2 = \{b^2\}, D(S_2, X_2) = \{b^3\}$.

References

[1] Berstel J. and D. Perrin, " Theory of Codes," Academic Press, Orlando, 1985

[2] Césari Y., *Sur un algorithme donnant les codes bipréfixes finis,* Math. Systems Theory **6**(1982), 221–225

[3] Choffrut C. and Culik II K., *On Extendibility of Unavoidable Sets,* Discrete Applied Mathematics **9**(1984), 125– 137

[4] Ehrenfeucht A. and G. Rozenberg, *Each Regular Code Is Included in a Regular Maximal Code,* RAIRO Informatique théorique **20**(1986), 89–96

[5] Eilenberg E., "Automata, Languages and Machines," Volume **A**, Academic Press, New York and London, 1974

[6] Higman G., *Ordering by Divisibility in Abstract Algebra,* Proc. London Math. Soc. **2(3)** (1952), 326–336

[7] Ito M., H. Jürgensen, H. J. Shyr and G. Thierrin, *Outfix and Infix Codes and Related Classes of Languages,* Journal of Computer and System Sciences **43**(1991), 484–508

[8] Lam N. H., *Finite Maximal Infix Codes,* (to appear in Semigroup Forum)

[9] Rosaz L., *Unavoidable Languaes, Cuts and Innocent Sets of Words,* Theoretical Informatics and Application **29**(1995), 339–382

[10] Shyr H. J., "Free Monoids and Languages," Lecture Notes, Hon Min Book Company, Taichung, 1991

[11] Shyr H. J. and G. Thierrin, *Codes and Binary Relations,* Lecture Notes 586 "Séminaire d'Algèbre, Paul Dubreil, Paris (1975–1976)," 180–188, Springer-Verlag

[12] Shyr H. J. and G. Thierrin, *Hypercodes,* Inform. Control **24**(1974), 45–54

[13] Van D. L., *Embedding Problem for Codes Defined by Binary Relations,* Preprint **98/A22**, Hanoi Institute of Mathematics

Strong recognition of rational ω-languages[1]

Bertrand Le Saëc, [2], V.R.Dare and R.Siromoney [3]

Abstract We propose a new notion of the recognition of w-languages using monoids. The used criteria is stronger than the one previously introduced by Büchi. It very easy to build deterministic automaton accepting a given rational ω-language from a monoid that strongly recognizes this w-language.

1 Introduction

Two notions of recognition used to characterize the rational langages of finite or infinite words has been intensively studied. The former is the recognition of langage using an automaton : a word is recognizes if its computation satisfies a given recognition criteria. The latter uses morphisms from the alphabet to monoid : a language is recognizes if its image in the monoid satisfies a given property. When the considered langage is a rational langage of finite words, it is easy to set a natural bijection between the deterministic automata recognizing a langage L and a set of right congruences of finite satisfying a property on their classes. In the same way, such a correspondance can be proposed between the monoids and the congruences of finite index. The property used for congruence and right congruence is a saturation one : if a word belongs to the language then the class of this word

[1]This work has been supported by the Indo-french Centre for the Promotion of Advanced Research (CEFIPRA)

[2]Laboratoire Bordelais de Recherche en Informatique – Unité de Recherche Associée au Centre National de la Recherche Scientifique n.1304, Université Bordeaux I 351, cours de la Libération 33405, Talence Cedex France

[3]Madras Christian College, TAMBRARAM, MADRAS 59, INDIA

must be included in the language. Since congruences are obviously right congruences, it is easy to obtain a deterministic automaton recognizing a langage from a monoid that recognizes this langage. From this result, it becomes simple to prove the Kleene theorem.

For rational ω-language (language of infinite words), Büchi introduced in [BU 62] a notion of recognition using monoids. This notion is based on a saturation property that can be easily "translate" in terms of congruences. There is many types of automata used to recognize rational languages. For the table transition automaton, a correspondance with right congruence has been also set in [LS 90], but the used saturation property is not of the same kind than the Büchi one.

In this note, we proposed a new definition of the recognition of rational ω-languages using monoids : *the strong monoid recoginition.* This notion is more powerful than those introduced by Büchi : when a monoid strongly recognizes a rational ω-language, it also recognizes the same ω-language in the Büchi sense, but the converse is false. We prove that a ω-language is rational iff it is strongly saturated by a finite monoid. Moreover, given a monoid which strongly recognizes a rational ω-language, the construction of a deterministic table transition automaton which accepts this ω-language is obvious. So a similar approach to those of the finite case can be made to set the Mac Naughton theorem : it consists to establish the equivalence of the Büchi recognition and the strong recognition. The sketch of the proof is : To build from a arbitrary automaton (Büchi or Müller or table transition) accepting a ω-language L a monoid that Büchi recognizes L, then to build from this monoid another one that strongly recognizes L and finally to construct a deterministic table transition automaton that accepts L.

Section 2 contains preliminary notions. The next section is devoted to the notion of strong monoid recognition. In the last section, we propose to study the Mac Naughton theorem by using strong recognition monoids.

2 Preliminaries

Let Σ be a finite alphabet. We denote by Σ^* and Σ^ω the sets of all finite and infinite words over Σ respectively. For any subset $X \neq \emptyset$ of Σ^*, we denote $X^+ = \{x_1x_2\cdots x_n,\ n \geq 1,\ x_i \in X\}$, $X^* = X^+ \cup \{\varepsilon\}$, where ε is the empty word, and $X^\omega = \{x_1x_2\cdots,\ \forall i \geq 1,\ x_i \in X \setminus \{\varepsilon\}\}$.

A language is a subset of Σ^* and a language of infinite words (ω-language) is a subset of Σ^ω. We denote by $UP(L)$ the set of its ultimately periodic words: $UP(L) = \{ xy^\omega \in L,\ x, y \in \Sigma^+\}$. An ω-language L is rational if it is a finite union of ω-languages of the form CB^ω where C, B are rational languages in Σ^*. In [BU 62], Büchi proved the following property : Let L, L' be rational ω-languages
then $L \neq \emptyset \iff UP(L) \neq \emptyset$ and so $L \subseteq L' \iff UP(L) \subseteq UP(L')$.

A transition system TS is a triple (Q, I, δ) where Q is a finite set of states, $I \subset Q$ is a set of initial states and δ is a transition function, *i.e.*, a partial mapping of $Q \times \Sigma$ into 2^Q. As usual, we extend δ to $Q \times \Sigma^*$ by setting, for all $q \in Q$, $a \in \Sigma$ and $w \in \Sigma^*$, $\delta(q,\varepsilon) = \{q\}$ and $\delta(q, aw) = \bigcup_{p\in\delta(q,a)} \delta(p,w)$. A transition system TS is deterministic if $Card(I) = 1$ and for any pair $(q,a) \in Q \times \Sigma$, there is at most one state q' in $\delta(q,a)$. By abuse, we will write $\delta(q,a) = q'$ when TS is deterministic.

Let $TS = (Q, I, \delta)$ be a transition system. A transition is an element (q,a,q') of $Q \times \Sigma \times Q$ such that $q' \in \delta(q,a)$. We denote by Δ the set of transitions of TS. A computation c in TS is a finite or infinite sequence $\delta_0\delta_1\delta_2\cdots$ of transitions where for each i, $\delta_i = (q_i, a_i, q_{i+1}) \in \Delta$. The word $w = a_0a_1\cdots$ is called the label of c and the state q_0 the origin of c. We use the notation $c(q_0, w, q_{n+1})$ to denote a finite computation c with the origin q_0, the end q_{n+1} and the label $w = a_0a_1\cdots a_n$ and for each i, $\delta_i = (q_i, a_i, q_{i+1}) \in \Delta$. We denote by $T\_fin(c)$ (resp. $T\_inf(c)$) the set of the transitions which have a finitely (resp. infinitely) many occurences in c.

Definitions 2.1 *A transition table automaton [LS 90] is a 4-tuple $\mathcal{A} = (Q, I, \delta, \mathcal{T})$ where (Q, I, δ) is a transition system and $\mathcal{T}$ a set of subsets of transitions. An infinite word $w \in \Sigma^\omega$ is accepted by $\mathcal{A}$ if there exists a computation c in A with the origin in I and the label w such that $T\_inf(c) \in \mathcal{T}$.*

The transition table automata accepts the rational ω-languages and their recognition power does not decrease if we only consider the deterministic ones [LS 90]. In the sequel, we denote $\mathcal{L}(\mathcal{A})$ the ω-langage accepted by an automaton $\mathcal{A}$.

Let $\equiv$ be an equivalence relation on Σ^*. We denote by $x_\equiv$ the class of the word x and we say that $\equiv$ is of finite index if it has a finite number of classes. A right congruence $\sim$ on Σ^* is an equivalence relation satisfying: $\forall u,\ v \in \Sigma^*$, $\forall w \in \Sigma^*$, $u \sim v$ implies $uw \sim vw$. A congruence $\approx$ on Σ^* is an equivalence relation satisfying: $\forall u,\ v \in \Sigma^*$, $\forall w,\ w' \in \Sigma^*$, $u \approx v$ implies $wuw' \approx wvw'$. As usual, with any congruence $\approx$ on Σ^*, we associate a monoid $M_\approx$ and a morphism $\varphi_\approx : \Sigma^* \longrightarrow M_\approx$ defined by $M_\approx = \{u_\approx, u \in \Sigma^*\}$ and for any $u \in \Sigma^*$, $\varphi(u) = u_\approx$. Symetrically, with any M is a monoid and any morphism $\varphi : \Sigma^* \longrightarrow M_\approx$, we associate a congruence $\approx_M$ defined by $\forall u, v \in \Sigma^*, u \approx_M v$ iff $\varphi(u) = \varphi(v)$. A stabilizer g of an element m of a monoid M satisfies the identity $mg = g$ and an idempotent f of M satisfies $f^2 = f$.

Definition 2.2 *[BU 62] A congruence $\approx$ on Σ^** Büchi-recognizes *an ω-language L iff for any pair m, f in $M_\approx$, we have $\varphi_\approx^{-1}(m)[\varphi_\approx^{-1}(f)]^\omega \cap L \neq \emptyset$ implies $\varphi_\approx^{-1}(m)[\varphi_\approx^{-1}(f)]^\omega \subset L$. If the congruence $\approx$ is of finite index, it is easy to show that the elements m, f can be choose in such a way that $m = mf$ and $f^2 = f$.*

Definitions 2.3 *[AR 85] Let $L \in \Sigma^\omega$ and let $\approx_s$ be the congruence on Σ^* defined by:*

$$\forall u,\ v \in \Sigma^*, u \approx_s v \Longleftrightarrow \begin{cases} \forall (x,w) \in \Sigma^* \times \Sigma^\omega,\ xuw \in L \Leftrightarrow xvw \in L \\ \forall x,\ w,\ w' \in \Sigma^* \text{ with } ww' \neq \varepsilon,\ x(wuw')^\omega \in L \Leftrightarrow x(wvw')^\omega \in L. \end{cases}$$

We denote by M_s the monoid associated with $\approx_s$. The congruence $\approx_s$ (resp. the monoid M_s, the morphism φ_s) is called the syntactic congruence (resp. monoid, morphism) of L. The congruence $\approx_s$ is the largest congruence of finite index which Büchi-recognizes a rational ω-language [AR 85].

There is a canonical bijection between the family of right congruences of finite index and the family of complete deterministic transition systems (up to isomorphism). This canonical bijection is defined in the following way:
With any right congruence of finite index $\sim$, we associate a deterministic transition system $TS_\sim = (Q_\sim, I_\sim, \delta_\sim)$ where $Q_\sim = \{x_\sim, x \in \Sigma^*\}$, $I_\sim = \{\varepsilon_\sim\}$, $\delta_\sim$ is the function from $Q_\sim \times \Sigma$ to $Q_\sim$ defined by $\forall (x_\sim, a) \in Q_\sim \times \Sigma$, $\delta_\sim(x_\sim, a) = (xa)_\sim$. With any deterministic transition system $TS = (Q, \{q_0\}, \delta)$, we associate a right congruence $\sim_Q$ defined by $\forall u, v \in \Sigma^*, u \sim_Q v$ iff $\delta(q_0, u) = \delta(q_0, v)$. The right congruence $\sim_Q$ is of finite index.
It is easy to see that the right congruence associated with $TS_\sim$ is exactly $\sim$ and, conversely, that the transition system associated with $\sim_Q$ is TS.

3 Strong congruence recognition

Definition 3.1 *A congruence $\approx$ (or its induced monoid $M_\approx$) on Σ^** strongly recognizes *an ω-language L if , for any triple $m,\ f,\ g \in M_\approx$ such that $mf = mg = m$, we have :*
$\varphi_\approx^{-1}(m)[(\varphi_\approx^{-1}(f))^+(\varphi_\approx^{-1}(g))^+]^\omega \cap L \neq \emptyset$ *implies* $\varphi_\approx^{-1}(m)[(\varphi_\approx^{-1}(f))^+(\varphi_\approx^{-1}(g))^+]^\omega \subset L$.

The following property is simple to set :

Lemma 3.2 *Let L be a rational ω-language accepted by a table transition automaton $\mathcal{A} = (Q, \{q_0\}, \delta, \mathcal{T})$. Let $\approx_Q$ be the relation defined by :*

$$\forall u,\ v \in \Sigma^*, u \approx_Q v \Longleftrightarrow \begin{cases} \forall q \in Q,\ \delta(q,u) = \delta(q,v) \\ \forall q, q' \in Q,\ T\_fin(c(q,u,q')) = T\_fin(c(q,v,q')) \end{cases}$$

Then $\approx_Q$ is a congruence of finite index which Büchi-recognizes L and so its induced monoid $M_{\approx_Q}$ also Büchi-recognizes L.

Notation 3.3 *Let φ be a morphism from Σ^* to a monoid M. Let m, f, g be three elements of M. We denote $\varphi^{-1}(m))[(\varphi^{-1}(f))^+(\varphi^{-1}(g))^+]^\omega$ by $F_{(\varphi,m,f,g)}$.*

Theorem 3.4 *An ω-language is rational iff it is strongly saturated by congruence of finite index.*

Proof : Let $\approx$ be a congruence of finite index that strongly recognizes an ω-language L. In order to prove that the ω-language L is rational, we can identify, in the definition of the strong recognition, the two elements f and g. This shows that a congruence which strongly recognizes a ω-langage L is also a congruence which Büchi-recognizes L. And so L is rational.

Conversely, let L be a rational ω-language accepted by a deterministic table transition automaton $\mathcal{A} = (Q, \{q_0\}, \delta, \mathcal{T})$. Let $\approx_Q$ the congruence defined in the previous lemma.

Now, let $m, f, g \in M_{\approx_Q}$ such that $m = mf = mg$. Since L is a rational ω-language, the languages $\varphi_{\approx_Q}^{-1}(m)$, $\varphi_{\approx_Q}^{-1}(f), \varphi_{\approx_Q}^{-1}(g)$ are rational and $F_{(\varphi_{\approx_Q}, m, f, g)}$: is a rational ω-language. So, in order to prove that $\approx_Q$ strongly saturates L, we just need to prove the following equivalence : $UP(F_{(\varphi_{\approx_Q}, m, f, g)} \cap L \neq \emptyset \iff UP(F_{(\varphi_{\approx_Q}, m, f, g)}) \subseteq L$

Since $m = mf = mg$, it is easy to see any ultimately periodic word of the ω-language $F_{(\varphi_{\approx_Q}, m, f, g)}$ is of the form $y(u_1^{p_1} v_1^{p'_1} \cdots u_k^{p_k} v_k^{p'_k})^\omega$ with $y \in \varphi_{\approx_Q}^{-1}(m)$ and for any $1 \leq i \leq k$, $u_i \in \varphi_{\approx_Q}^{-1}(f), v_i \in \varphi_{\approx_Q}^{-1}(g)$. Moreover, $\mathcal{A}$ is deterministic and for any $1 \leq i \leq k$, $y \approx_Q yu_i \approx_Q yv_i$ so there exists $q \in Q$ such that $\delta(q_0, y) = q$, $\delta(q, u_i) = q$ and $\delta(q, v_i) = q$.

For any $u \in \varphi_{\approx_Q}^{-1}(f)$ and $v \in \varphi_{\approx_Q}^{-1}(g)$, we have $\forall k \leq 0, T\_fin(c(q, u^k, q)) = T\_fin(c(q, u, q))$ and $\forall k \leq 0, T\_fin(c(q, v^k, q)) = T\_fin(c(q, v, q))$ so $T\_fin(c(q, u_1^{p_1} v_1^{p'_1} \cdots u_k^{p_k} v_k^{p'_k}, q)) = T\_fin(c(q, uv, q))$ where u, v are two fixed words of $\varphi_{\approx_Q}^{-1}(f)$ and $\varphi_{\approx_Q}^{-1}(g)$ respectively. So, we have $T\_fin(c(q, u_1^{p_1} v_1^{p'_1} \cdots u_k^{p_k} v_k^{p'_k}, q)) \in \mathcal{T} \iff T\_fin(c(q, uv, q)) \in \mathcal{T}$. Let x be an arbitrary word in $x \in \varphi_{\approx_Q}^{-1}(m)$, we have : $UP(F_{(\varphi_{\approx_Q}, m, f, g)} \cap L) \neq \emptyset \iff x(uv)^\omega \in L$ and $UP(F_{(\varphi_{\approx_Q}, m, f, g)}) \subseteq L$. So $\approx_Q$ strongly recognizes L. $\square$

Remark 3.5 *In the proof of the previous Theorem, it is shown that any congruence which strongly recognizes an ω-langage L is also a congruence which Büchi-recognizes the same ω-language. The converse is false.*

For instance, let $L = \{a, b\}^*(ab)^\omega$. The class of the syntactic congruence of the word aa also contains the two words aaa and aab so we have $aa \approx_s aaa \approx_s aab$. Let m_{aa} (resp. m_a, m_b) denote the element the syntactic monoid of L corresponding to aa (resp. (a, b)). Since $aa(ab)^\omega \in L$, we have $\varphi_{\approx_s}^{-1}(m_{aa})[(\varphi_{\approx_s}^{-1}(m_a))^+(\varphi_{\approx_s}^{-1}(m_b))^+]^\omega) \cap L \neq \emptyset$; but $\varphi_{\approx_s}^{-1}(m_{aa})[(\varphi_{\approx_s}^{-1}(m_a))^+(\varphi_{\approx_s}^{-1}(m_b))^+]^\omega$ is obviously not included in L since $aa(abb)^\omega \notin L$. The syntactic congruence of L Büchi recognizes L, but it does not strongly recognizes L.

Definition 3.6 *Let $\sim$ be a congruence of finite index. Let $\mathcal{A}_\sim$ be the deterministic table transition automaton $\mathcal{A}_\sim = (Q_\sim, I_\sim, \delta_\sim, \mathcal{T}_\sim)$ where $(Q_\sim, I_\sim, \delta_\sim) = TS_\sim$ is the transition system inced by $\sim$ and $\mathcal{T}_\sim = \{T\_inf(\varepsilon_\sim, w)$ such that $w \in UP(L)\}$. This automaton is called the deterministic table transition automaton induced by $\sim$.*

Proposition 3.7 *Let $\approx$ be a congruence that strongly recognizes a rational ω-language L then $\mathcal{A}_\approx$*

is a deterministic table transition that accepts L.

Proof. Let L be a rational ω-language and $\approx$ be a congruence that strongly recognizes L. In [LS 90], it has been proved that if a right congruence of finite index $\sim$ recognizes L in the following way :

1. $\forall (u,v) \in \Sigma^* \times \Sigma^*, \forall w \in \Sigma^\omega :\ u \sim v$ implies $\{uw \in L \iff vw \in L\}$
2. $\forall (x,u,v) \in \Sigma^* \times \Sigma^* \times \Sigma^*,\ x \sim xu \sim xv$ implies $\{x(u^+v^+)^\omega \cap L \neq \emptyset \implies x(u^+v^+)^\omega \subseteq L\}$.

then the deterministic table transition automaton induced by $\sim$ accepts the ω-language L. So it is suffisant to show that $\approx$ recognizes L : in this sense.

1. Let $u,v \in \Sigma^*$ such that $u \approx v$ and let $w \in \Sigma^\omega$ such that $uw \in L$. There exists $m, m', f \in M_\approx$ such that $u \in \varphi_\approx^{-1}(m), w' \in \varphi_\approx^{-1}(m'), w'' \in \varphi_\approx^{-1}(f)^\omega$ and $w = w'w''$. So $\varphi_\approx^{-1}(mm')\varphi_\approx^{-1}(f)^\omega \cap L \neq \emptyset$. Moreover $\approx$ strongly recognizes L, so it also Büchi-recognizes L and $\varphi_\approx^{-1}(mm')\varphi_\approx^{-1}(f)^\omega \subseteq L$. Since $u \approx v$, we have $vw' \in \varphi_\approx^{-1}(mm')$ and finally $vw'w'' = vw \in L$.
2. Let $x \in \Sigma^*$ and $u,v \in \Sigma^+$ such that $x \approx xu \approx xv$ and $x(u^+v^+)^\omega \cap L \neq \emptyset$. Let $m, g, f \in M_\approx$ such that $m = \varphi_\approx(x), g = \varphi_\approx(u)$ and $f = \varphi_\approx(v)$. Then $F_{(\varphi,m,f,g)} \cap L \neq \emptyset \implies F_{(\varphi,m,f,g)} \subset L$ and so $x(u^+v^+)^\omega \subseteq L$.

Finally, the automaton $\mathcal{A}_\approx$ accepts L. □

4 Mac Naughton Theorem and strong recognition

The previous theorem provides a direct way to prove the Mac Naughton theorem : we "just" need to build a congruence of finite index that stongly recognizes a rational ω-language. We propose two manners to obtain such a proof. The first one can be view as an intrepretation in term of monoids of the proof proposed in [LPW 92]. The second is based on a conjecture concerning the role of commutativity for the stabilizers in a monoid.

4.1 Nice semigroups

The two following results has been set in [LPW 92] :

Theorem 4.1 *Any finite monoid M is the image by a surjective morphism of a finite monoid $\hat{M}$ in which the stabilizers satisfy the identities $f^2 = f$ and $fgf = fg$.*

Proposition 4.2 *An ω-language L is rational iff it is Büchi-recognized by a finite nice semigroup.*

The proof of the previous result consists to start with a finite semigroup M that Büchi-recognizes L, to build the corresponding finite nice semigroup $\hat{M}$ and to easily show that $\hat{M}$ also Büchi-recognizes L.

Proposition 4.3 *A finite nice semigroup that Büchi-recognizes a rational ω-language L also strongly recognizes L*

Proof. Let L be a rational ω-language and M be a finite nice semigroup that Büchi-recognizes L. So there exist a morphism $\varphi: \Sigma^* \longrightarrow M$ such that for any pair m, f in M, we have $\varphi^{-1}(m)[\varphi^{-1}(f)]^\omega \cap L \neq \emptyset$ implies $\varphi^{-1}(m)[\varphi^{-1}(f)]^\omega \subset L$.

Let $m, f, g \in M$ such that $mf = mg = m$ and $F_{(\varphi,m,f,g)} \cap L \neq \emptyset$.
Let $x \in \varphi^{-1}(m), u \in \varphi^{-1}(f), v \in \varphi^{-1}(g)$ be three arbitrary words.

Since M is a nice semigroup, the stabilizers of M satisfy $f^2 = f$ and $fgf = fg$. In fact, in the sequel of the proof, we just need one of the two properties : $f^2 = f$.

Since f and g are idempotents, we have $\varphi^{-1}(f)^+ = \varphi^{-1}(f)$, and $\varphi^{-1}(g)^+ = \varphi^{-1}(g)$ so $F_{(\varphi,m,f,g)} = \varphi^{-1}(m)[(\varphi^{-1}(f))(\varphi^{-1}(g))]^\omega$.

Let $y(u_1v_1 \cdots u_kv_k)^\omega \in UP(F_{(\varphi,m,f,g)} \cap L)$ with $y \in \varphi^{-1}(m)$ and $\forall 1 \leq i \leq k$, $u_i \in \varphi^{-1}(f), v_i \in \varphi^{-1}(g)$.

Since M Büchi-recognizes L, it is larger than the syntactic monoid of L. So $\forall 1 \leq i \leq k$, $u \approx_s u_1 \approx_s u_2 \cdots \approx_s u_k$ and $v \approx_s v_1 \approx_s v_2 \cdots \approx_s v_k$. So, we have $y(u_1v_1 \cdots u_kv_k)^\omega \in F_{(\varphi,m,f,g)} \cap L \neq \emptyset \Longleftrightarrow x(uv)^\omega \in F_{(\varphi,m,f,g)} \cap L$. Finally : $F_{(m,f,g)} \cap L \neq \emptyset \Longleftrightarrow F_{(\varphi,m,f,g)} \subseteq L$.
□

4.2 Semigroups with commutative stabilizers

A semigroups with commutative stabilizers is a semigroup in which the stabilizers satisfy the identity $fg = gf$.

Proposition 4.4 *A finite semigroup with commutative stabilizers that Büchi-recognizes a rational ω-language L also strongly recognizes L*

Proof. Let L be a rational ω-language and M be a finite semigroup with commutative stabilizers that Büchi-recognizes L. So there exist a morphism $\varphi: \Sigma^* \longrightarrow M$ such that for any pair m, f in M, we have $\varphi^{-1}(m)[\varphi^{-1}(f)]^\omega \cap L \neq \emptyset$ implies $\varphi^{-1}(m)[\varphi^{-1}(f)]^\omega \subset L$.

Let $m, f, g \in M$ such that $mf = mg = m$ and $F_{(\varphi,m,f,g)} \cap L \neq \emptyset$.
Let $x \in \varphi^{-1}(m)$, $u \in \varphi^{-1}(f)$, $v \in \varphi^{-1}(g)$ be three arbitrary words.

Let $y(u_1v_1 \cdots u_kv_k)^\omega \in UP(F_{(\varphi,m,f,g)} \cap L)$ with $y \in \varphi^{-1}(m)$ and $\forall 1 \leq i \leq k$, $u_i \in \varphi^{-1}(f), v_i \in \varphi^{-1}(g)$.

Since M Büchi-recognizes L, it is larger than the syntactic monoid of L so $\forall\, 1 \leq i \leq k$, we have $u \approx_s u_1 \approx_s u_2 \cdots \approx_s u_k$ and $v \approx_s v_1 \approx_s v_2 \cdots \approx_s v_k$ then $w \in UP(F_{(\varphi,m,f,g)} \cap L) \Longleftrightarrow y(u^{p_1}v^{p'_1} \cdots u^{p_k}v^{p'_k})^\omega \in UP(F_{(\varphi,m,f,g)} \cap L)$.

Moreover, f, g satisfy the identity $fg = gf$, we have :
$w \in UP(F_{(\varphi,m,f,g)} \cap L) \Longleftrightarrow y(u^{p_1+\cdots+p_k}v^{p'_1+\cdots+p'_k})^\omega \in UP(F_{(\varphi,m,f,g)} \cap L)$.
Now, M is finite and there exists $\alpha > 0$ such that u^α and v^α are idempotent of M. So
$y(u^{p_1+\cdots+p_k}v^{p'_1+\cdots+p'_k})^\omega \in UP(F_{(\varphi,m,f,g)} \cap L) \Longleftrightarrow$
$y((u^{p_1+\cdots+p_k}v^{p'_1+\cdots+p'_k})^\alpha)^\omega \in UP(F_{(m,f,g)} \cap L) \Longleftrightarrow$
$y(u^{(p_1+\cdots+p_k)*\alpha}v^{(p'_1+\cdots+p'_k)*\alpha})^\omega \in UP(F_{(\varphi,m,f,g)} \cap L \Longleftrightarrow$
$y(u^\alpha v^\alpha)^\omega \in UP(F_{(\varphi,m,f,g)} \cap L \Longleftrightarrow UP(F_{(\varphi,m,f,g)} \subset eqL$ □

Remark 4.5 *We don't know if a result similar to the theorem 3.1 can be set replacing "nice semigroup" by "semigroup with commutative stabilizers". Our initition is based on the fact that around a state in an automaton the loops can commutes without changing the accepting language. But if it is true this should lead to an another proof of the Mac Naughton theorem.*

References

[AR 85] A. Arnold *"A syntactic congruence for rational ω-languages"* Theor. Comp. Sci. 39 (1985) 333-335.

[BU 62] J.R Büchi *"On a decision method in restricted second-order arithmetic"* In E. Nagel, P. Suppes and A Tarski editors, "Logic, Methodology and Philosophy of Sciences: Proc of the 1960 International Congress", Stanford University Press, (1962) 1-11.

[LS 90] B. Le Saëc *"Saturating right congruences"* R.A.I.R.O. Inform. Theor. Appl. 24(6) (1990) 545-560.

[LPW 92] B. Le Saëc, J.E. Pin, P. Weil, *"Semigroups with idempotent stabilizers and applications to automata theory"*, International Journal of Algebra and Computation, Vol 1(3) (1991) 291-314.

[MN 66] R. McNaughton *"Testing and generating infinite sequences by finite automaton."* Inform. Control 9, (1966) 521-30.

SOME RESULTS CONCERNING COVERS IN THE CLASS OF MULTIVALUED POSITIVE BOOLEAN DEPENDENCIES

LE DUC MINH AND VU NGOC LOAN

Vietnam National University (VNU)

NGUYEN XUAN HUY

Institute of Information Technology
National Center for Natural Science and Technology of Vietnam

In this paper we present a new class of data dependencies that called multivalued positive Boolean dependencies. This class is a generalization of all classes of positive Boolean dependencies. Some generalization concepts for implications as well as an axiom system for inference rules are also presented. With the help of the equivalence theorem of consequences in the world of all relations, the world of 2-tuple relations and propositional logic in the class of multivalued positive Boolean dependencies we can get some significant results related to m-covers. Some conditions for two sets being m-equivalent are given. Necessary and sufficient conditions for testing an m-redundant set of multivalued positive Boolean dependencies are introduced. Base on these results some algorithms for constructing a nonredundant m-cover of a given set of multivalued positive Boolean dependencies are presented.

1. Introduction

An important part of the design of relational database schema is the specification of constraints or data dependencies in databases. In other words, central to the design of database schemes is the idea of a data dependency, that is, a constraint on the possible relations that can be the current instance of a relation scheme.

Data dependencies represent a semantic tool for expresssing properties of data and play an important role to ensure consistency of data. There are many kinds of data dependencies that have been studied, such as, functional dependencies, strong dependencies, equational dependencies, generalizied positive Boolean dependencies, etc. [1, 2, 3, 4, 8].

In this paper we intend to mention a new class of data dependencies that called *Multivalued positive Boolean dependencies.* This class is a generalization of all classes of positive Boolean dependencies [8, 9, 10, 11]. In this class we can also see data of objects managed in database in fuzzy semantics. This is useful and significant for other purposes when using databases.

Some generalization concepts for implications as well as an axiom system for inference rules are also presented. With the help of the equivalence theorem of consequences in the world of all relations, the world of 2-tuple relations and propositional logic in the class of multivalued positive Boolean dependencies we can get some significant results.

In fact, we want to find, for a given set of dependencies, an equivalent set with a number of useful properties and a shorter representation of constraints. For several algorithms with running times that are dependent on the number of dependencies in the input, a smaller set of dependencies guarantees faster execution. Dependencies are used in database systems to help ensuring consistency and corrections. Fewer dependencies mean less storage space used and fewer tests to make when the database is modified.

Also with the help of the equivalence theorem we can get some significant results related to m-covers. Necessary and sufficient conditions for testing an m-redundant set of multivalued positive Boolean dependencies are introduced. On the basis of these results, some algorithms for constructing a nonredundant m-cover of a given set of multivalued positive Boolean dependencies are presented.

2. Some concepts and results

Some basic concepts concerning with the class of multivalued positive Boolean dependencies have been mentioned in [9]. Let U be a set of attributes, $U = \{A_1, A_2, ..., A_n\}$. For each attribute A_i in U, beside the domain d_i we have a set B_i called the valuation domain of A_i that satisfies the following conditions:

(i) $B_i \subset [0, 1]$,

(ii) $1 \in B_i$, and

(iii) If $s \in B_i$ then $1 - s \in B_i$.

We denote $\boldsymbol{B} = B_1 \times B_2 \times ... \times B_n$.

Definition 1. *A formula f over U is said to be a multi-valued positive Boolean dependency (MVPBD) iff $f(e) = 1$ with $e = (1, 1, ..., 1) \in \boldsymbol{B}$. By MF_p we denote the set of all multivalued positive Boolean dependencies over U.*

Definition 2. *For each domain d_i of attribute A_i, $1 \leq i \leq n$, we consider a mapping $\alpha_i : d_i \times d_i \to B_i$, satisfying the following :*

(i) $(\forall a \in d_i)\ (\alpha_i(a, a) = 1)$,
(ii) $(\forall a, b \in d_i)\ (\alpha_i(a, b) = \alpha_i(b, a))$, *and*
(iii) $(\forall s \in B_i, \exists a, b \in d_i)\ (\alpha_i(a, b) = s)$

For each $k \in [0, 1]$, $f \in MF_p$ and $\Sigma \subseteq MF_p$, we set $T_f^k = \{x \in \boldsymbol{B} \mid f(x) \geq k\}$ and $T_\Sigma^k = \{x \in \boldsymbol{B} \mid \forall f \in \Sigma, f(x) \geq k\}$.

By $REL(U)$ we denote the set of all relations over U. Let $R \in REL(U)$ and $u, v \in R$. Then $(\alpha_1(u.A_1, v.A_1), ..., \alpha_n(u.A_n, v.A_n))$ is said to be a valuation over U and denoted by $\alpha(u, v)$, where $u.A_i$ is the value of attribute A_i in tuple u. We set $T_R = \{\alpha(u, v)/u, v \in R\}$. Note that for every $u \in R$ we have $\alpha(u, u) = e \in T_R$.

Suppose β is a mapping $\beta : [0, 1] \to [0, 1]$, then β is called *a level.*

Definition 3. *Let $R \in REL(U)$, $k \in [0, 1]$, $f \in MF_p$ and $\Sigma \subseteq MF_p$, $m \in [0, 1]$ and $\beta : [0, 1] \to [0, 1]$. Then*

(a) *We say that R k-satisfies f, iff $T_R \subseteq T_f^k$. R is said to k-satisfy the set Σ iff $T_R \subseteq T_\Sigma^k$. If R k-satisfies Σ then we denote $R^k(\Sigma)$. Instead of $R^k(\{f\})$ we write $R^k(f)$.*
(b) *We say that R (β, m)-satisfies f, iff R $\beta(m)$-satisfies f. R is said to (β, m)-satisfy the set Σ iff R $\beta(m)$-satisfies Σ.*

Definition 4. *Let $\Sigma \subseteq MF_p$, $f \in MF_p$, $m \in [0, 1]$ and β be a level*

(a) *We say that f is implied from Σ by relations and write $\Sigma \overset{m}{\models} f$ iff for any relation R m-satisfying Σ then R also (β, m)-satisfies f*
(b) *If for any relation R having two tuples and R m-satisfies Σ we also have R (β, m)-satisfies f then we denote $\Sigma \overset{m}{\models}_2 f$.*

We set $\Sigma_m^* = \{f \mid \Sigma \overset{m}{\models} f\}$.

Definition 5. *Let $\Sigma \subseteq MF_p$, $f, g \in MF_p$, $m \in [0,1]$ and β is a level. We consider a system of two rules for deductions $|\frac{m}{\beta}$ as follows:*

(a) If $f \equiv 1$ then $\Sigma |\frac{m}{\beta} f$,
(b) If $\Sigma |\frac{m}{\beta} f$ and $T_\Sigma^m \cap T_f^{\beta(m)} = T_\Sigma^m \cap T_g^{\beta(m)}$ then $\Sigma |\frac{m}{\beta} g$.

Definition 6. *Suppose $m \in [0,1]$, $\Sigma \subseteq MF_p$, $f \in MF_p$ and β is a level. We say that Σ (β, m)-implies f or f is (β, m)-implied from Σ, denoted by $\Sigma |\frac{m}{\beta} f$ iff f can be obtained from Σ after a finite number of steps applying rul es of the above axiom system. For brief, we write $\Sigma |\!\!\frac{m}{\quad} f$ instead of $\Sigma |\frac{m}{\beta} f$.*

We set $\Sigma_m^+ = \{f \,|\, \Sigma |\!\!\frac{m}{\quad} f\}$.

Definition 7. *A set $\Sigma \subseteq MF_p$ is said to be stable iff $\Sigma \subseteq \Sigma_m^+$.*

Note that if $\beta(m) = m$ then any set Σ is stable.

Proposition 1. *[11] Suppose that $\Sigma \subseteq MF_p$, $f \in MF_p$ and β is a level, then the following are equivalent:*

(a) $\Sigma |\!\!\frac{m}{\quad} f$
(b) $f(x) \geq \beta(m)$ for any $x \in T_\Sigma^m$
(c) $T_\Sigma^m \subseteq T_f^{\beta(m)}$

Proposition 2. *If $\Sigma, \Gamma \subseteq MF_p$ and $\Sigma \supseteq \Gamma$ then*

(a) $T_\Sigma^m \subseteq T_\Gamma^m$
(b) $\Sigma_m^+ \supseteq \Gamma_m^+$
(c) If $\Gamma |\!\!\frac{m}{\quad} f$ then $\Sigma |\!\!\frac{m}{\quad} f$.

Proof. It is clear that $T_\Sigma^m \subseteq T_\Gamma^m$. Now suppose $f \in \Gamma_m^+$. It means $\Gamma |\!\!\frac{m}{\quad} f$. Then $f(x) \geq \beta(m)$ for any $x \in T_\Gamma^m$. Hence, if $x \in T_\Sigma^m$ then we also have $f(x) \geq \beta(m)$. It means that $\Sigma |\!\!\frac{m}{\quad} f$. We get $f \in \Sigma_m^+$. From b we get the assertion c. The proof is complete. □

Proposition 3. *Let $m \in [0,1]$ with $\beta(m) \geq m$ and $\Sigma \subseteq MF_p$. Then $\Sigma_m^+ \supseteq \Sigma_m^{++}$.*

Proof. We set $\Gamma = \Sigma_m^+$, so $\Sigma_m^{++} = \Gamma_m^+$. For any $f \in \Sigma_m^{++}$ we get $f \in \Gamma_m^+$. By using Proposition 1 we obtain $T_\Gamma^m \subseteq T_f^{\beta(m)}$ (1). Because of $\Sigma \mid\!\overset{m}{\text{—}} \Gamma$, $\beta(m) \geq m$ and Proposition 1 we have $T_\Sigma^m \subseteq T_\Gamma^{\beta(m)} \subseteq T_\Gamma^m$ (2). From (1) and (2) we obtain $T_\Sigma^m \subseteq T_f^{\beta(m)}$. According to Proposition 1 we have $\Sigma \mid\!\overset{m}{\text{—}} f$ and hence $\Sigma_m^+ \supseteq \Sigma_m^{++}$. □

Proposition 4. *Let $m \in [0,1]$ with $\beta(m) \geq m$ and $\Sigma, \Gamma \subseteq MF_p$. If $\Sigma \mid\!\overset{m}{\text{—}} \Gamma$ then $\Sigma_m^+ \supseteq \Gamma_m^+$.*

Proof. Suppose $f \in \Gamma_m^+$ (1), we have to show that $f \in \Sigma_m^+$ (2). From (1) and Proposition 1 we get $T_\Gamma^m \subseteq T_f^{\beta(m)}$ (3). From the hypothesics $\Sigma \mid\!\overset{m}{\text{—}} \Gamma$, Proposition 1 and $\beta(m) \geq m$ we get $T_\Sigma^m \subseteq T_\Gamma^{\beta(m)} \subseteq T_\Gamma^m$ (4). Combining (3) and (4) we obtain $T_\Sigma^m \subseteq T_f^{\beta(m)}$. It implies that assertion (2) holds and therefore we have $\Sigma_m^+ \supseteq \Gamma_m^+$. The proof is complete. □

It is not hard to see the following propositions

Proposition 5. *If $\Sigma, \Gamma \subseteq MF_p$ such that Γ is stable and $\Sigma_m^+ \supseteq \Gamma_m^+$ then $\Sigma \mid\!\overset{m}{\text{—}} \Gamma$.*

Proposition 6. *Let $m \in [0,1]$ with $\beta(m) \geq m$ and let $\Sigma, \Gamma \subseteq MF_p$ such that Σ, Γ are stable. Then $\Sigma_m^+ = \Gamma_m^+$ iff $\Sigma \mid\!\overset{m}{\text{—}} \Gamma$ and $\Gamma \mid\!\overset{m}{\text{—}} \Sigma$.*

Proposition 7. *Let $m \in [0,1]$ with $\beta(m) \geq m$ and $\Sigma, \Gamma, \Delta \subseteq MF_p$. If $\Sigma \mid\!\overset{m}{\text{—}} \Gamma$ and $\Gamma \mid\!\overset{m}{\text{—}} \Delta$ then $\Sigma \mid\!\overset{m}{\text{—}} \Delta$.*

Proposition 8. *Let $m \in [0,1]$ with $\beta(m) \geq m$. Suppose that $\Sigma \subseteq MF_p$ and Σ is stable then $\Sigma_m^{++} = \Sigma_m^+$.*

Definition 8. *Suppose $m \in [0,1]$. Two sets $\Sigma, \Gamma \subseteq MF_p$ are said to be m-equivalent and denoted by $\Sigma \overset{m}{=} \Gamma$ iff $\Sigma_m^* = \Gamma_m^*$.*

Theorem 1. (Equivalence Theorem [11]). *Let $m \in [0,1]$, $\Sigma \subseteq MF_p$, $f \in MF_p$ and β be a level. Then the following are equivalent:*

(a) $\Sigma \mid\!\overset{m}{\text{—}} f$

(b) $\Sigma \overset{m}{\models} f$

(c) $\Sigma \overset{m}{\models}_2 f$

Corollary 1. *Two sets* Σ, $\Gamma \subseteq MF_p$ *are m-equivalent iff* $\Sigma_m^+ = \Gamma_m^+$.

Proposition 9. *Let* $m \in [0,1]$ *and* $\Sigma, \Gamma \subseteq MF_p$. *If* $T_\Sigma^m = T_\Gamma^m$ *then two sets* Σ *and* Γ *are m-equivalent.*

Let $R \in REL(U)$, $f \in MF_p$ and let $m \in [0,1]$. Now we consider an algorithm for checking whether a relation R satisfying a given dependency. We have known that for $f \in MF_p$, $\Sigma \subseteq MF_p$ and $m \in [0,1]$ then the followings hold:

(a) $R^m(f)$ if and only if $T_R \subseteq T_f^m$

(b) $R^m(\Sigma)$ if and only if $T_R \subseteq T_\Sigma^m$.

Algorithm SATISFIES

Input - A relation R over U,

- A formula f,
- A real number m in [0,1]

Output - TRUE if $R^m(f)$

- FALSE, otherwise

Format SATISFIES (R, f, m)

Method

// *Verify whether the assertion* $T_R \subseteq T_f^m$ *is valid*

return $(T_R \subseteq T_f^m)$;

End SATISFIES.

Definition 9. *Suppose* $m \in [0,1]$ *and* $\Sigma, \Gamma \subseteq MF_p$. *If* Σ *and* Γ *are m-equivalent then we say that* Γ *is a m-cover of* Σ, *and conversely* Σ *is a m-cover of* Γ.

Definition 10. *Let* $m \in [0,1]$, $\Sigma \subseteq MF_p$ *and* $g \in \Sigma$. *We say that g is m-redundant in* Σ *iff* $\Sigma - \{g\} \overset{m}{\models} g$. *The set* Σ *is said to be m-redundant iff there exists g in* Σ *such that g is m-redundant in* Σ.

Proposition 10. *Let* $m \in [0,1]$ *with* $\beta(m) \geq m$, $\Sigma \subseteq MF_p$ *and let* $g \in \Sigma$ *and g be m-redundant in* Σ. *Then two sets* Σ *and* $\Sigma - \{g\}$ *are m-equivalent.*

Proof. We set $\Gamma = \Sigma - \{g\}$. Because of $\Sigma \supseteq \Gamma$ and Proposition 2 we have

$$\Sigma_m^+ \supseteq \Gamma_m^+ \tag{1}$$

On the other hand, for $f \in \Sigma_m^+$, we have to show that $f \in \Gamma_m^+$. Suppose $x \in T_\Gamma^m$. Because of g being m-redundant in Σ and Equivalence Theorem we have $\Gamma \mid\!\frac{m}{\;}\, g$ and $g(x) \geq \beta(m) \geq m$. Hence $x \in T_\Sigma^m$ and $f(x) \geq \beta(m)$. It means $f \in \Gamma_m^+$. We obtain

$$\Sigma_m^+ \subseteq \Gamma_m^+ \tag{2}$$

From (1), (2) and Corollary 1 we see that two sets Σ and Γ are m-equivalent□

Here we have a necessary and sufficient condition for a dependency g beging m-redudent in Σ.

Proposition 11. *Let $\Sigma \subseteq MF_p$, $m \in [0,1]$, β be a level and let $g \in \Sigma$. Then g is m-redundant in Σ iff $T^m_{\Sigma-\{g\}} \subseteq T_g^{\beta(m)}$.*

Proof. Suppose that a dependency g is m-redundant in Σ. By Definition 10 we have $\Sigma - \{g\} \models^{m} g$. Using Equivalence Theorem we obtain $\Sigma - \{g\} \mid\!\frac{m}{\;}\, g$. This is also equivalent to $T^m_{\Sigma-\{g\}} \subseteq T_g^{\beta(m)}$. Conversely, suppose that $T^m_{\Sigma-\{g\}} \subseteq T_g^{\beta(m)}$. According to Proposition 1 we have $\Sigma - \{g\} \mid\!\frac{m}{\;}\, g$. Also using Equivalence Theorem we obtain $\Sigma - \{g\} \models^{m} g$. It means g being m-redundant. □

Corollary 2. *Let $\Sigma \subseteq MF_p$, $m \in [0,1]$ and β be a level. Then Σ is m-redundant iff there exists $g \in \Sigma$ such that $T^m_{\Sigma-\{g\}} \subseteq T_g^{\beta(m)}$.*

Proof. Suppose that there exists $g \in \Sigma$ such that $T^m_{\Sigma-\{g\}} \subseteq T_g^{\beta(m)}$. By applying Proposition 1 and Theorem 1 we see that g is m-redundant in Σ. Hence Σ is m-redundant.

Conversely, suppose that Σ is m-redundant. Hence there exists $g \in \Sigma$ such that $\Sigma - \{g\} \models^{m} g$. Using Equivalence Theorem we have $\Sigma - \{g\} \mid\!\frac{m}{\;}\, g$. According to Propossition 1 we obtain $T^m_{\Sigma-\{g\}} \subseteq T_g^{\beta(m)}$. □

Corollary 3. *Let $\Sigma \subseteq MF_p$, $m \in [0,1]$ and β be a level. Suppose that there exists two formulas $f, g \in \Sigma$ such that $f \neq g$ and $T_f^m \subseteq T_g^{\beta(m)}$. Then Σ is m-redundant.*

Proof. Set $\Gamma = \Sigma - \{g\}$. Then $f \in \Gamma$ and therefore $T_\Gamma^m \subseteq T_f^m$. Combining with the hypothesis $T_f^m \subseteq T_g^{\beta(m)}$ we get $T_\Gamma^m \subseteq T_g^{\beta(m)}$. Using Corollary 1 we can see that Σ is m-redundant. □

Definition 11. *Let $m \in [0,1]$. A set $\Sigma \subseteq MF_p$ is said to be m-nonredundant if Σ is not a m-redundant set of formulas.*

Now we consider some algorithms concerning with m-redundant sets.

Algorithm MEMBER

Input - A subset Σ of MF_p

- A formula f in Σ
- A real number m in [0, 1]

Output - TRUE, if $\Sigma \overset{m}{\models} f$,

- FALSE, otherwise

Format MEMBER(Σ, f, m)

Method

return $(T_\Sigma^m \subseteq T_f^{\beta(m)})$;

end Member.

Algorithm DERIVES

Input - Two sets Σ and Γ of MF_p

- A real number m in [0,1]

Output - TRUE, if $\Sigma \overset{m}{\models} \Gamma$

- FALSE, otherwise

Format DERIVES(Σ, Γ, m);

Method // *Using Equivalence Theorem.*

```
    for each g in Γ do
        if not MEMBER(Σ, g, m) then
            return FALSE
        endif;
    endfor;
    return TRUE;
```

end DERIVES.

Algorithm EQUIVALENCE

Input - Two subsets Σ and Γ of MF_p such that they are stable

- A real number $m \in [0,1]$ with $\beta(m) \geq m$

Output - TRUE, if $\Sigma \overset{m}{=} \Gamma$

- FALSE, otherwise

Format EQUIVALENCE(Σ, Γ, m)

Method // *Using Proposition 6 and Corollary 1*

return (DERIVES(Σ, Γ, m) **and** DERIVES(Γ, Σ, m)

end EQUIVALENCE.

Algorithm REDUNDANT

Input - A subset Σ of MF_p

- A real number m in [0,1], β is a level

Output - TRUE, if Σ is m-redundant

- FALSE, otherwise

Format REDUNDANT(Σ, m)

Method // *Using Corollary 2*

for each g **in** Σ **do**

if $T^m_{\Sigma-\{g\}} \subseteq T^{\beta(m)}_g$ **then**

return TRUE

endif;

endfor;

return FALSE;

end REDUNDANT.

Algorithm NONREDUN

Input - A subset Σ of MF_p

- A real number m in [0,1], β is a level

Output - An m-nonredundant cover of Σ

Format NONREDUN(Σ, m)

Method // *Step by step we delete dependencies which are m-redundant*

// *in Σ such that the rest set of dependencies is also a m-cover of Σ*

```
  Γ := Σ;
  for each g in Σ do
      if MEMBER(Γ − {g}, g, m) then
          Γ := Γ − {g}
      endif;
  endfor;
  return Γ;
end NONREDUN.
```

We consider the case $\beta(m) = m$. We will have some concepts and results as follows [9].

Definition 12. *Let $\Sigma \subseteq MF_p$ and $f \in MF_p$. We denote*

(a) $\Sigma \models f$ *iff for any $m \in [0,1]$ the assertion* $\Sigma \overset{m}{\models} f$ *holds*
(b) $\Sigma(x) = \bigwedge_{f \in \Sigma} f(x)$

Theorem 2. *Suppose $\Sigma \subseteq MF_p$ and $f \in MF_p$. Then $\Sigma \models f$ iff $\Sigma(x) \leq f(x)$ for any $x \in \boldsymbol{B}$.*

Definition 13. *Let $\Sigma, \Gamma \subseteq MF_p$. We say that Σ and Γ are equivalent and write $\Sigma \cong \Gamma$ iff $\Sigma \overset{m}{=} \Gamma$ for any $m \in [0,1]$.*

Corollary 4. *Suppose $\Sigma, \Gamma \subseteq MF_p$. Then $\Sigma \cong \Gamma$ iff $\Sigma(x) = \Gamma(x)$ for any $x \in \boldsymbol{B}$.*

Definition 14. *Let $\Sigma \subseteq MF_p$. A dependency g in Σ is said to be redundant iff $\Sigma \cong \Sigma - \{g\}$. If there exists $h \in \Sigma$ such that h is redundant then Σ is also called to be redundant.*

Corollary 5. *Assume that $\Sigma \subseteq MF_p$ and $g \in \Sigma$. Then g is redundant in Σ iff for any $m \in [0,1]$ g is m-redundant in Σ.*

Proof. (a) *Necessary condition.* Suppose $g \in \Sigma$ and g is redundant in Σ. We set $\Gamma = \Sigma - \{g\}$. Then we have $\Sigma \cong \Gamma$. Therefore $\Gamma \models g$. Applying Definition

12 it follows that for any $m \in [0,1]$ we always obtain $\Gamma \overset{m}{\models} g$. It means that g is m-redundant in Σ.

(b) *Sufficient condition.* Let $g \in \Sigma$ and suppose that for any $m \in [0,1]$ then g is m-redundant in Σ. According to the hypothesis, we have $\Sigma - \{g\} \overset{m}{\models} g$. Because of m is arbitrary and $m \in [0,1]$ it follows that $\Sigma - \{g\} \models g$. Therefore g is a redundant dependency in Σ. Sufficient condition has been proved and this completes the proof. □

Corollary 6. *Let $g \in \Sigma$ and let $\Gamma = \Sigma - \{g\}$. Then the dependency g is redundant in Σ iff $\Gamma(x) \leq g(x)$ for any $x \in \boldsymbol{B}$.*

Acknowledgments

We would like to thank Prof. Ho Thuan for carefully reading earlier versions of the paper. We also thank Prof. Masami Ito for helpful comments.

References

1. Beeri C., On the Membership Problem for Fuctional and Multivalued Dependencies in Relational Databases. *ACM* TODS 5, 3, September 1980, 241-259.
2. Berman J. and Blok W. J., Generalized Boolean dependencies. *Abstracts of AMS*, 6 (1985), 163.
3. Berman J. and Blok W. J., Positive Boolean dependencies. *Inf. Processing Letters*, 27(1988), 147-150.
4. Codd E. F., A Relational Model of Data for Large Shared Data Banks *CACM* 13 : 6 June 1970, 377-387.
5. Date C. J., *An introduction to Databases System.* Additon Wesley Publishing Company, London, 1982.
6. Ho Thuan, Some invariants of covers for functional dependencies. *MTA SZTAKI Kozlemeayek* 34/1986.
7. Maier D. J., Minimum covers in the relational database model. *J ACM 27* (Oct. 1980), 664-674.
8. Nguyen Xuan Huy and Le Thi Thanh, Generalized Positive Boolean Dependecies. *J. Inform. Process. Cybernet. EIK* 28 (1992) 6, 363-370.
9. Vu Ngoc Loan, Nguyen Xuan Huy, A class of generalized logical dependencies in deductive databases. *Vietnam Fourth Informatics Week, Proceeding*, 195-203.

10. Vu Ngoc Loan, Le Duc Minh, The class of extended functional dependencies in the relational data model. *VNU Journal of Science. Nat. Sci.*, T. XIII, No. 2, 1977, p. 1-5.
11. Le Duc Minh, Vu Ngoc Loan, Nguyen Xuan Huy, Some results concerning with the class of multivalued positive boolean dependencies in the relational data model in context of fuzzy semantics. *Proceedings of VJ-FUSSSY '98 : Vietnam - Japan bilaterial Symposium on Fuzzy Systems and Applications*, Ha Long Bay, Vietnam, 30th September - 2nd October, 1998, p. 378-382.

A NEW MEASURE FOR ATTRIBUTE SELECTION

Do Tan Phong
Mobi Fone Company VMS, Hanoi, Vietnam
Ho Thuan
Institute of Information Technology IIT, Hanoi, Vietnam
Ha Quang Thuy
College of Sciences, VNU

Abstract. In this article, we propose a new measure for attribute selection (R_N - measure) having closed relations to rough measure (Pawlak Z. [6]) and R - measure (Ho Tu Bao, Nguyen Trung Dung [3]) . We prove that all of these three measures are confidence measures i.e. satisfy the weak monotonous axiom. So the R_N - measure is worth in the class of attribute selection measures. Some relations between these three measures are also shown.

Key words: confidence measure, rough measure, R-measure, R_N -measure.

1. The weak monotonous axiom

Following Dubois D. and Prade H. [1], measures in approximate reasoning should satisfy the weak monotonous axiom. The weak monotonous axiom of a measure is described as follows.

Let Ω be a set (Ω is called by reference set) and let g be a positive function defined on the subsets of Ω ($g : 2^\Omega \to R; \forall A \subseteq \Omega$, we have $g(A) \geqslant 0$). A measure g is called to satisfy *the weak monotonous axiom* (in this article, it is called by *the monotonous axiom*) if:

$$\forall A, B \subseteq \Omega : A \subseteq B \Rightarrow g(A) \leqslant g(B) \quad (1)$$

The monotonous axiom is one of main requirements that measures in approximate reasoning should be have. The meaning of it may be explained as the follows: By having more information for reasoning then having more belief. The axiom should be checked when we create a new measure in approximate reasoning. A measure which satisfies the monotonous axiom is called by a *confidence measure.*

2. Measures for attribute selection

The data gathered from difference sources almost are rough data and relations between these data almost are unknown. These data usually described in form of two- dimensions table, where a row is the data of an object, and a column is the data of an attribute. One of the relationship should be considered is the attribute dependency: Exists or not a relation between a group of attributes and another group of attributes and how to determine the value of these relations? The determination of dependency between some groups of attributes is one of main problems in analysis, discovery relations of data systems. Measures for attribute selection are defined to solve above problems.

Definition 1. Let O be a set of objects and $E \subseteq O \times O$ is an equivalence relation on O. Two objects $o_1, o_2 \in O$ are said to be distinguished by E if they satisfied equivalence relation E (or $o_1 E o_2$).

Definition 2. Let O be a set of objects and $E \subseteq O \times O$ is a equivalence relation on $O, X \subseteq O$. Then sets $E_*(X)$ and $E^*(X)$ are defined as following:

$$E_*(X) = \{o \in O | [o]_E \subseteq X\} \tag{2}$$

$$E^*(X) = \{o \in O | [o]_E \cap X \neq \emptyset\} \tag{3}$$

(where$[o]_E$ denoted the equivalent class consisting of objects distinguished with o by the equivalence relation E). $E_*(X)$ and $E^*(X)$ are respectively called by the lower approximation and upper approximation of X.

Lower approximation and upper approximation, defined by above definition show out an approximation of set X through distribution set of objects of X by an equivalence relation.. Some contents about lower approximation and upper approximation sets are deal in [2,3,5,6,7]. Let Ω be a set of attributes, P be a subset of Ω. P determines one equivalence relation on the set of objects O and partitions O into equivalent classes, each class consists of all objects that have the same value for all attributes in P.

Let P and Q, two subset of Ω. P and Q will partition O into different equivalent classes and when we consider relations between equivalent classes by these two partitions, we have some information of causal relations from P and Q. This information is explained in form of measures for attributes selection [3].

3. The measure R_N

	Temperature	Headache	flu
E_1	Normal	Yes	No
E_2	High	Yes	Yes
E_3	Very high	Yes	Yes
E_4	Normal	No	No
E_5	High	No	No
E_6	Very high	No	Yes
E_7	High	No	No
E_8	Very high	Yes	Yes

Table 1. Table of collect data

Measures for attributes selection by definitions 3 and 4 have been described in [3,6]. To explain some contents in this article, we use the data in table 1 (with the assumption that there do not exist two rows with same values [3]):

Definition 3. (Pawlak Z. [6]) Let O be a set of objects, let Ω be set of attributes, $P, Q \subseteq \Omega$ are two subset of attributes. Then the rough measure (is denoted by$\mu_P(Q)$), measuring the dependence level of the subset Q on subset P is determined as follow:

$$\mu_P(Q) = \frac{card(\{o \in O | [o]_P \subset [o]_Q\})}{card(O)} \tag{4}$$

Definition 4. (Ho Tu Bao, Nguyen Trung Dung [3]) Let O be a set of objects, let Ω be set of attributes, $P, Q \subseteq \Omega$ are two subset of attributes. Then the measure R (is denoted by $\widetilde{\mu}_P(Q)$) measures the dependence level of the subset Q on subset P is determined as follow:

$$\widetilde{\mu}_P(Q) = \frac{1}{card(O)} \sum_{[o]_P} \max[o]_Q \frac{card^2([o]_P \cap [o]_P)}{card([o]_P)} \tag{5}$$

As corresponding to the data in Table 1, the dependence degree of the ***flu*** attribute on the ***headache*** by (5) equal to ***9/16*** while the correlative rough measure by (4) equal to ***0***.

In the following, we will propose a new confidence measure, the measure R_N, which is smaller than "the possibility measure" R and is greater than the rough measure.

Definition 5. Let O be a set of objects, letΩ be set of attributes, $P, Q \subset \Omega$ are two subset of attributes. Then the measure R_N (is denoted by μ_P^N) which measures the dependence level of the subset Q on subset P is determined as follow:

$$\mu_P^N(Q) = \frac{1}{card(O)} \Big(\sum_{[o]_P \subseteq [o]_Q} + \sum_{[o]_P \not\subseteq [o]_Q} \sum_{[o]_Q} \frac{card^2([o]_Q \cap [o]_P)}{card^2([o]_P)} \Big). \tag{6}$$

As corresponding to the data in Table 1, the dependence degree of the ***flu*** attribute on the ***headache*** by (6) equal to ***5/32***.

4. Some characteristics of the measure R_N

Proposition 1. *Let Ω be set of attributes, $P, Q \subseteq \Omega$ are two subset of attributes. Then:*

$$\mu_P(Q) \leqslant \mu_P^N(Q) \leqslant \tilde{\mu}_P(Q)$$

Proof:

At first, we write the formulas of rough measure and measure R by another form as follows:

$$\mu_P(Q) = \frac{card(\{o \in O | [o]_P \subseteq [o]_Q)\}}{card(O)} = \frac{1}{card(O)} \Big(\sum_{[o]_P \subseteq [o]_Q} card([o]_P) \Big) \tag{a}$$

Let

$$A = \frac{1}{card(O)}\Big(\sum_{[o]_P\subseteq[o]_Q} \max[o]_Q \frac{card^2([o]_Q\cap[o]_P)}{card([o]_P)}\Big)$$

Consider $[o]_P \subseteq [o]_Q$, we would like to show that

$$\max[o]_Q \frac{card^2([o]_Q\cap[o]_P)}{card([o]_P)} = card([o]_P). \tag{b}$$

Because $[o]_P \subseteq [o]_Q$ then $\exists![o]_Q$ satisfying $[o]_Q\cap[o]_P \neq \oslash$ and the maximum on equivalent classes $[o]_Q$ reaches on this $[o]_Q$. By other hand, we have $card([o]_Q \cap [o]_P) = card([o]_P)$ and (b) is confirmed. That

$$\tilde{\mu}_P(Q) = \mu_P(Q) + \frac{1}{card(O)}\Big(\sum_{[o]_P\not\subseteq[o]_Q} \max[o]_Q \frac{card^2([o]_Q\cap[o]_P)}{card([o]_P)}\Big) \tag{c}$$

- By the definition 5 and (a), we have $\mu_P(Q) \leqslant \mu_P^N(Q)$.
- We prove now the second inequality $\mu_P^N(Q) \leqslant \tilde{\mu}_P(Q)$

By the definition 6 and (c), we have only to prove:

$$\sum_{[o]_P\not\subseteq[o]_Q}\sum_{[o]_Q} \frac{card^2([o]_Q\cap[o]_P)}{card^2([o]_P)} \leqslant \sum_{[o]_P\not\subseteq[o]_Q} \max_{[o]_Q} \frac{card^2([o]_Q\cap[o]_P)}{card([o]_P)} \tag{d}$$

We consider for each class $[o]_P$ in case there is no class $[o]_Q$ which contained it.

Denote

$$B = \sum_{[o]_Q} \frac{card^2([o]_Q\cap[o]_P)}{card^2([o]_P)} = \frac{1}{card^2([o]_P)}\sum_{[o]_Q} card^2([o]_Q\cap[o]_P).$$

Because the number of positive components which belong to B is not exceed the number of objects in $[o]_P$ (it means that card $\{[o]_Q : [o]_Q\cap[o]_P \neq \oslash\} \leqslant card([o]_P)$), the number of non-zero terms which belong to the sum is not exceed the cardinal of $[o]_P$. For each term, we have:

$$card^2([o]Q\cap[o]P) \leqslant \max_{[o]_Q}(card^2([o]_Q\cap[o]_P))$$

and then

$$\begin{aligned} B &\leqslant \frac{1}{card^2([o]_P)} card([o]_P) \max_{([o]_Q)}(card^2([o]_Q\cap[o]_P)) \\ &= \frac{1}{card([o]_P)} \max_{([o]_Q)}(card^2([o]_Q\cap[o]_P)). \end{aligned}$$

Thus, the corresponding components in two hands of (d) satisfy the inequality, and (d) is proved so $\mu_P^N(Q) \leqslant \tilde{\mu}_P(Q)$.

Let Ω be set of attributes, $P, Q \subseteq \Omega$ are two subset of attributes. When considering the dependence of subset Q on subset P, then P is called by the condition attribute set and Q is called by the decision attribute set.

For rules in form "if P then Q", the belief of them depends on the change of parameters P and Q. In the following, we investigate the belief of this rule on direction, in which we fixed the decision parameter Q and changed the condition parameter P.

Proposition 2. *Let Ω be set of attributes. $\forall P, Q \subseteq \Omega$. We have $\tilde{\mu}_P(Q) \leqslant 1$.*

Proof:

$$\forall P, Q \subseteq \Omega, \forall o \in O \text{then}([o]_P) \cap [o]_Q \subseteq [o]_P, \forall [o]_Q$$
$$\Leftrightarrow \max_{[o]_Q} \frac{card^2([o]_Q \cap [o]_P)}{card([o]_P)} \leqslant \frac{card^2([o]_P)}{card([o]_P)} = card([o]_P)$$
$$\Leftrightarrow \tilde{\mu}_P(Q) = \frac{1}{card(O)} \sum_{[o]_P} \max_{[o]_Q} \frac{card^2([o]_Q \cap [o]_P)}{card([o]_P)} \leqslant \frac{1}{card(O)} \sum_{[o]_P} card([o]_P) = 1$$

Proposition 3. *Let O be set of objects. For two subsets of attributes P, Q we have:*
$$\forall o \in O, [o]_P \subseteq [o]_Q \text{iff} \mu P(Q) = \mu_P^N(Q) = \tilde{\mu}_P(Q) = 1.$$

Proof: For the rough measure (Pawlak), the equation in the Proposition is obvious.
By propositions 1 and 2, we have: $\mu_P(Q) \leqslant \mu_P^N(Q) \leqslant \tilde{\mu}_P(Q) \leqslant 1$
$\Rightarrow \forall o \in O, [o]_P \subseteq [o]_Q \Leftrightarrow 1 = \mu_P(Q) \leqslant \mu_P^N(Q) \leqslant \tilde{\mu}_P(Q) \leqslant 1$
$\Rightarrow \forall o \in O, [o]_P \subseteq [o]_Q \Leftrightarrow 1 = \mu_P(Q) = \mu_P^N(Q) = \tilde{\mu}_P(Q) = 1$

Corollary 1. *Let Ω be set of attributes, $\forall Q \subseteq \Omega$, we have*

$$\mu_\Omega(Q) = \mu_\Omega^N(Q) = \tilde{\mu}_\Omega(Q) = 1.$$

Proposition 4. *$\forall P, Q \subseteq \Omega, (P \cap Q) = \oslash$, and $\bar{P}$ is denoted the compensate set of P in Ω, we have:*
$$\mu_{\bar{P}}(Q) = \mu_{\bar{P}}^N(Q) = \tilde{\mu}_{\bar{P}}(Q) = 1.$$

Proof: By proposition 3.

Remarks 1. *$\forall$ integers a_i, b_i, where a_i are no negative, b_i positive $(i = 1, 2, ..., n)$, we have*
$$\frac{\left(\sum_{i=1}^n a_i\right)^2}{\sum_{i=1}^n b_i} \leqslant \sum_{i=1}^n \frac{a_i^2}{b_i} \text{and} \frac{\left(\sum_{i=1}^n a_i\right)^2}{\sum_{i=1}^n b_i} \leqslant \sum_{i=1}^n \frac{a_i^2}{b_i^2}$$

Theorem 1. *The rough measure of Pawlak, the measure R, the measure R_N satisfy the monotonous axiom.*

Proof: Consider two subset of attributes P and P' where $P \subseteq P'$. Let m be any one of three above measures, we should prove that $m(P') \geqslant m(P)$.

Preliminary remarks:

Assume that the set of objects O is partitioned by the subset of attributes P into q equivalent classes. Because $P \subseteq P'$ then each i-equivalent class by the subset of attributes P consists $n_i (i = 1, 2, ..., q)$ equivalent classes by the subset of attributes P'.

Denote the representative object of j-equivalent class $(j = 1, 2, ...n_i)$ by the subset of attributes P' which is contained the i-equivalent class by the subset of attributes P is $o_i^j (j = 1, 2, ..., n_i)$. For each i-equivalent class by the subset of attributes P, we denote the representative object is o_i^*. We chose the object o_i^* be one of the objects o_i^j in some cases.

We consider i-equivalent class (it means that $[o_i^*]_P$) by the subset of attributes P, we have:

(*) $[o_i]_P = \bigcup_{j=1}^{n_i} [o_i^j]'_P$

(**) $card([o_i]_P) = \bigcup_{j=1}^{n_i} card([o_i^j]'_P)$

(***) For each equivalent class $[o]_Q$ by the subset of attributes Q: $card([o]_Q \cap [o_i]_P) = \bigcup_{j=1}^{n_i} card([o]_Q \cap [o_i^j]'_P)$

• Let m be the rough measure:

Consider two sets $O_1 = \{o \in O : [o]_P \subseteq [o]_Q\}$ and $O_2 = \{o \in O : [o]'_P \subseteq [o]_Q]\}$

For each $o \in O_1$, we consider the equivalence class $[o]_P$. By above, we have o be some o_i and $[o_i]'_P \subseteq [o_i]_P \subseteq [o_i]_Q$, that is $o \in O_2$. Because in spite of o so $O_1 \subseteq O_2$.

Since then $card(O_1) \leqslant card(O_2)$ whether $m(P) \leqslant m(P')$.

• Let m be the measure R:

According to the preliminary remarks, we have following inequalities:

$$card(O) \times \tilde{\mu}_P(Q) = \sum_{[o]_P} \max_{[o]_Q} \frac{card^2([o]_Q \cap [o]_P)}{card([o]_P)} = \sum_{i-1}^{q} \max_{[o]_Q} \frac{card^2([o]_Q \cap [o_i^*]_P)}{card([o_i^*]_P)}$$

and

$$card(O) \times \tilde{\mu}_{P'}(Q) = \sum_{[o]_{P'}} \max_{[o]_Q} \frac{card^2([o]_Q \cap [o]_{P'})}{card([o]_{P'})} = \sum_{i-1}^{q} \sum_{i=1}^{n_i} \max_{[o]_Q} \frac{card^2([o]_Q \cap [o_i^j]_{P'})}{card([o_i^j]_{P'})}$$

Because of $card(O)$ fixed so for proving $\tilde{\mu}_P(Q) \leqslant \tilde{\mu}_{P'}(Q)$ we have only to prove that:

$$\sum_{i-1}^{q} \max_{[o]_Q} \frac{card^2([o]_Q \cap [o_i^*]_P)}{card([o_i^*]_P)} \leqslant \sum_{i=1}^{q} \sum_{j=1}^{n_i} \max_{[o]_Q} \frac{card^2([o]_Q \cap [o_i^j]_{P'})}{card([o_i^j]_{P'})} \tag{e}$$

Two members of (e) have q of terms, so to confirm this inequality, we only confirm for each corresponding terms in q couples of terms. That is for $i = 1, 2, .., q$, we prove that:

$$\max_{[o]_Q} \frac{card^2([o]_Q \cap [o_i^*]_P)}{card([o_i^*]_P)} \leqslant \sum_{j=1}^{n_i} \max_{[o]_Q} \frac{card^2([o]_Q \cap [o_i^j]_{P'})}{card([o_i^j]_{P'})} \tag{f}$$

According to the preliminary remarks, we may chose o_i^* be representative object of the i-equivalence class by subset of attributes P with some special properties. We chose o_i^* be the object which maximizes $\frac{card^2([o]_Q \cap [o_i^*]_P)}{card([o_i^*]_P)}$ (it belongs to the equivalent class by subset of attributes Q, too). By this choise o_i^* belongs to $[o]_P$ and since $[o]_P$ is partitioned into classes $[o_i^j]_{P'}, o_i^*$ belongs to some j_o- equivalent class: the equivalent class $[o_i^{j_o}]_{P'}$. We chose the representative object $o_i^* = o_i^{j_o}$ which have equivalent class by subset Q ($[o_i^{j_o}]_Q$) which maximizes the left member of (f). Since then, the left member of (f) is equal to

$$\frac{card^2([o_i^{j_o}]_Q \cap [o_i^{j_o}]_P)}{card([o_i^{j_o}]_P)} \tag{g}$$

For the right member of (f), for $j = 1, 2, ..., n_i$, we always have:

$$\max_{[o]_Q} \frac{card^2([o]_Q \cap [o_i^j]_{P'})}{card([o_i^j]_{P'})} \geqslant \frac{card^2([o_i^{j_o}]_Q \cap [o_i^j]_{P'})}{card([o_i^j]_{P'})}$$

so that

$$\sum_{j=1}^{n_i} \max_{[o]_Q} \frac{card^2([o]_Q \cap [o_i^j]_{P'})}{card([o_i^j]_{P'})} \geqslant \sum_{j=1}^{n_i} \frac{card^2([o_i^{j_o}]_Q \cap [o_i^j]_{P'})}{card([o_i^j]_{P'})} = B$$

According to the first inequality of the ***Remarks 1***, we have:

$$B \geqslant \frac{\left(\sum_{j=1}^{n_i} card([o_i^{j_o}]_Q \cap [o_i^j]_{P'})\right)^2}{\sum_{j=1}^{n_i} card([o_i^j]_{P'})} = \frac{card^2([o_i^{j_o}]_Q \cap [o_i^{j_o}]_P)}{card([o_i^{j_o}]_P)}$$

(by equality (**) and (***) in the preliminary remarks and we chose $o_i^{j_o}$ to be the representative object o_i^* in equivalent class by P).

So that

$$\sum_{j=1}^{n_i} \max_{[o]_Q} \frac{card^2([o]_Q \cap [o_i^j]_{P'})}{card([o_i^j]_{P'})} \geqslant \frac{card^2([o_i^{j_o}]_Q \cap [o_i^{j_o}]_P)}{card([o_i^{j_o}]_P)}$$

Since then, (f) is proved for any i-term ($i = 1, 2, .., q$) that is $m(P) \leqslant m(P')$ or by another term, $R(P) \leqslant R(P')$.

- Let m be the measure R_N:

Like above, we consider:

$$\begin{aligned} card(O) \times \mu_P^N(Q) &= \sum_{[o]_P \subseteq [o]_Q} card([o]_P) + \sum_{[o]_P \not\subseteq [o]_Q} \sum_{[o]_Q} \frac{card^2([o]_Q \cap [o]_P)}{card^2([o]_P)} \\ &= A + \sum_{[o]_P \not\subseteq [o]_Q} \sum_{[o]_Q} \frac{card^2([o]_Q \cap [o]_P)}{card^2([o]_P)} \end{aligned}$$

in there $A = \sum_{[o]_P \subseteq [o]_Q} card([o]_P)$ and

$$card(O) \times \mu_{P'}^N(Q) = \sum_{[o]_{P'} \subseteq [o]_Q} card([o]_{P'}) + \sum_{[o]_{P'} \not\subseteq [o]_Q} \sum_{[o]_Q} \frac{card^2([o]_Q \cap [o]_{P'})}{card^2([o]_{P'})}$$

Due to $card(O)$ is fixed so to prove $\mu_P^N(Q) \leqslant \mu_{P'}^N(Q)$ we prove the inequality for only two right members of the formulas.

We have:

$$\forall [o]_P : \sum_{[o]_Q} \frac{card^2([o]_Q \cap [o]_P)}{card^2([o]_P)} \leqslant card[o]_P \qquad \text{(h)}$$

because of $card([o]_P) = \sum_{[o]_Q} card([o]_Q \cap [o]_P)$ and

$$\frac{card^2([o]_Q \cap [o]_P)}{card^2([o]_P)} = card([o]_Q \cap [o]_P)\frac{card([o]_Q \cap [o]_P)}{card^2([o]_P)} \leqslant card([o]_Q \cap [o]_P)$$

We classify equivalent classes by subset of attributes P' into three following kinds:

+ $[o]_{P'} \subseteq [o]_P \subseteq [o]_Q$ are equivalent classes by P' which are partitioned from equivalent classes by P such that $[o]_P \subseteq [o]_Q$. For every equivalent class attended in the sum A by partition of P is related with a group of equivalent classes attended in the sum A by partition of P' in this kind and two sums have the same value. We denote the set of equivalent classes $[o]_{P'}$ in this kind by set I.

+ $[o]_{P'} \subseteq [o]_Q$ but $[o]_{P'} \not\subseteq [o]_P$. The value of the term for this equivalent class by subset P' equal to $card([o]_{P'})$. We denote the set of equivalent classes $[o]_{P'}$ in this kind by set II.

+ $[o]_{P'} \not\subseteq [o]_Q$: We denote the set of equivalent classes $[o]_{P'}$ in this kind by set III.

To apply to the preliminary remarks and with out loss of generality the comprehensive we assume that equivalent classes by subset P which are related with equivalent classes by subset P' belonging to set I be the first classes $[o_i^*]_P (i = 1, 2, ..., k;$ for $0 \leqslant k \leqslant q)$. To observe that when $k = 0$ the there is not any equivalent class by subset P, that is contained in an equivalent class by subset Q; Otherwise, when $k = q$, all of equivalent classes by subset P are contained in some equivalent class by subset Q. We perform $card(O) \times \mu_P^N(Q)$ by another form:

$$\begin{aligned} card(O) \times \mu_P^N(Q) &= \sum_{i=1}^{k} card([o_i^*]_P) + \sum_{i=k+1}^{q} \sum_{[o]_Q} \frac{card^2([o]_Q \cap [o_i^*]_P)}{card^2([o_i^*]_P)} \\ &= A + \sum_{i=k+1}^{q} \sum_{[o]_Q} \frac{card^2([o]_Q \cap [o_i^*]_P)}{card^2([o_i^*]_P)} \qquad \text{(i)} \end{aligned}$$

We consider the following sum which related with subset P':

$$\begin{aligned}
card(O) \times \mu^N_{P'}(Q) &= \sum_{[o]_{P'} \subseteq [o]_Q} card([o]_{P'}) + \sum_{[o]_{P'} \not\subseteq [o]_Q} \sum_{[o]_Q} \frac{card^2([o]_Q \cap [o]_{P'})}{card^2([o]_{P'})} \\
&= \sum_{[o]_{P'} \in I} card([o]_{P'}) + \sum_{[o]_{P'} \in II} card([o]_{P'}) + \sum_{[o]_{P'} \in III} \sum_{[o]_Q} \frac{card^2([o]_Q \cap [o]_{P'})}{card^2([o]_{P'})} \\
&= \sum_{i=1}^{k} card([o_i^*]_P) + \sum_{[o]_{P'} \in II} card([o]_{P'}) + \sum_{[o]_{P'} \in III} \sum_{[o]_Q} \frac{card^2([o]_Q \cap [o]_{P'})}{card^2([o]_{P'})} \\
&= A + \sum_{[o]_{P'} \in II} card([o]_{P'}) + \sum_{[o]_{P'} \in III} \sum_{[o]_Q} \frac{card^2([o]_Q \cap [o]_{P'})}{card^2([o]_{P'})} \\
(\text{by (h)}) &\geqslant A + \sum_{[o]_{P'} \in II \cup III} \sum_{[o]_Q} \frac{card^2([o]_Q \cap [o]_{P'})}{card^2([o]_{P'})} = A + C \qquad \text{(j)}
\end{aligned}$$

with

$$C = \sum_{[o]_{P'} \in II \cup III} \sum_{[o]_Q} \frac{card^2([o]_Q \cap [o]_{P'})}{card^2([o]_{P'})}$$

After gathering all of equivalent classes by subset P' into equivalent classes by subset P, we get:

$$C = \sum_{i=k+1}^{q} \sum_{j=1}^{n_i} \sum_{[o]_Q} \frac{card^2([o]_Q \cap [o_i^j]_{P'})}{card^2([o_i^j]_{P'})} = \sum_{i=k+1}^{q} \sum_{[o]_Q} \sum_{j=1}^{n_i} \frac{card^2([o]_Q \cap [o_i^j]_{P'})}{card^2([o_i^j]_{P'})}$$

By the second inequality in the ***Remarks 1*** and the preliminary remarks (**), and (***), we have:

$$\sum_{j=1}^{n_i} [o]_Q \frac{card^2([o]_Q \cap [o_i^j]_{P'})}{card^2([o_i^j]_{P'})} \geqslant \frac{\sum_{j=1}^{n_i} card([o]_Q \cap [o_i^j]_{P'})^2}{\left(\sum_{j=1}^{n_i} card([o_i^j]_{P'})\right)^2} = \frac{card^2([o]_Q \cap [o_i^*]_P)}{card^2([o_i^*]_P)}$$

Then

$$C \geqslant \sum_{i=k+1}^{q} \sum_{[o]_Q} \frac{card^2([o]_Q \cap [o_i^*]_P)}{card^2([o_i^*]_P)} \qquad \text{(k)}$$

According (i), (j), and (k) we have $R_N(P) \leqslant R_N(P')$

By the corollary 1 and the theorem 1, we find out: if considering the set of all of attributes Ω be the reference set then the rough measure of Pawlak, the measure R, the measure R_N are confidence measures.

5. Conclusion

By Dubois D. and Prade H. [1], one pair of dual confidence measures are considered as two bound measures: The necessity measure N and the possibility measure Π. The necessity measure N is seemed as the minimum measure and the possibility measure Π is seemed as the maximum measure. Between two measures there is a set of confidence measures and probability is one of them. We may consider the measure R and the rough measure is two bound measures and R_N is one measure which lays between them (The proposition 1). However, the measure R and the rough measure have not some special relations as two measure Π and N.

References

1. Dubois Didier, Prade Henri (1986). *Possibility Theory: An Approach to Computerized Processing of Uncertainly.* CNRS, Languages and Computer Systems (LSI), University of Toulouse III (English edition: The University of Cambridge. 1988).
2. Ha Quang Thuy (1996). *Rough set in decision tables.*, Journal of Science, The Vietnam National University, Hanoi. Vol. 12. No 4-1996, pp 9-14 (in Vietnamese).
3. Ho Tu Bao, Nguyen Trong Dung (1996). *A Rough Sets Based Measure for Attribute Selection in Decision Tree Induction.* The Fourth International Workshop on Rough Sets, Fuzzy Sets and Machine Discovery (RSFD '96).
4. Jiawei Han and Yangjian Fu (1996). *Explorating of Attribute-Oriented Induction in Data Mining. In the book Advances in Knowledge discovery and Data mining.* AAAI Press / The MIT Press, 1996, pp 399-425.
5. Le Tien Vuong, Ho Thuan (1989). *A relation database extended by applications of fuzzy set theory and linguistic variables.* Computers and artificial Intelligence, Vol. 9, No.2, 153-168, 1989, Bratislava.
6. Pawlak Z. (1985). *Rough set and Decision Tables.* ICS PAS Report, 540, 3-1984, Warsawa, Poland.
7. Theresa Beaubouef, Frederik E. Petry, Gurdial Arora (1998). *Information-theoretic measures of uncertainty for rough sets and rough relational databases.* Journal of information Sciences. No 409 (1998) pp. 185-195.
8. Usama M. Fayyad, Gregory Piatetsky - Shapiro, Padhraic Smyth (1996). *From Data Mining to Knowledge Discovery: An Overview.* In the book *Advances in Knowledge discovery and Data mining.* AAAI Press / The MIT Press, 1996, pp 1-36.

The Complexity of Problems Defined by Boolean Circuits

Steffen Reith   Klaus W. Wagner
Lehrstuhl für Theoretische Informatik
Universität Würzburg
[streit,wagner]@informatik.uni-wuerzburg.de

Abstract

We study the complexity of circuit-based combinatorial problems (e.g., the circuit value problem and the satisfiability problem) defined by Boolean circuits with gates from an arbitrary finite base B of Boolean functions. Special cases have been investigated in the literature. We give a complete characterization of their complexity depending on the base B. For example, for the satisfiability problem for Boolean circuits with gates from B we present a complete collection of (decidable) criteria which tell us for which B this problem is in L, is complete for NL, is complete for $\oplus$L, is complete for P, or is complete for NP. Our proofs make substantial use of the characterization of all closed classes of Boolean functions given by E.L. POST already in the twenties of the last century.

1 Introduction

The complexity of formula-based and circuit-based combinatorial problems was studied through the more than three decades of Complexity Theory. Already in 1971, S.A. COOK [Coo71] proved that the satisfiability problem for Boolean formulae is NP-complete: This was the first NP-complete problem ever discovered. R.E. LADNER [Lad77] proved in 1977 that the circuit value problem is P-complete. In many cases when a new complexity class was introduced and investigated, a formula-based or circuit-based combinatorial problem was the first which was proved to be complete for this class (see [SM73, Gil77], for example). However, usually these problems were defined using formulae or circuits with a *complete base* of Boolean operations or gates, mostly with the base $\{\wedge, \vee, \neg\}$. But what about the complexity of such problems when a different base is used? There are several special results of this kind (e.g. [Sim75, Gol77, Lew79, GP86]). In particular, there are very detailed investigations for the special case of Boolean formulae in conjunctive normal form in [Sch78, Cre95, CH96, CH97, KST97, RV00]. But there are no results answering this question for unresricted circuits in full generality. In this paper we will give complete characterizations of the complexity of some combinatorial problems defined by circuits with Boolean gates from an arbitrary finite base.

The paper is organized as follows. In Section 2 we define B-circuits as Boolean circuits with gates from a finite set B of Boolean functions, and we define $\mathrm{BF}[B]$ as the class of Boolean functions computed by B-circuits. The complexity of a problem defined by B-circuits only depends on $\mathrm{BF}[B]$. The classes of the form $\mathrm{BF}[B]$ are exactly those classes of Boolean functions which contain the identity function and which are closed under superposition (i.e., substitution, permutation of variables, identification of variables, and introduction of non-essential variables). Already in the twenties, E.L. POST [Pos41] gave a complete characterization of these classes. We make substantial use of his results which we present in Section 3. In Sections 4-8 we study the complexity of the circuit value problem, the satisfiability problem and the tautology problem, some quantified circuit problems, the counting function, and the threshold problem, resp., when defined by circuits with Boolean gates from an arbitrary finite base B. We give complete characterizations of their complexity in terms of completeness for suitable complexity classes.

2 Problems Defined by B-Circuits

In this paper we will study problems which are defined by Boolean circuits with gates from a finite set of Boolean functions. Informally, for a finite set B of Boolean functions, a *B-circuit C with input variables* $x_1, \ldots, x_{\alpha(C)}$ is a directed acyclic graph with a special output node which has the following properties: Every vertex (gate) with indegree 0 is labeled with an x_i or a 0-ary function from B. Every vertex (gate) with indegree $k > 0$ is labeled with a k-ary function from B. Given values $a_1, \ldots, a_{\alpha(C)} \in \{0, 1\}$ to $x_1, \ldots, x_{\alpha(C)}$, every gate computes a Boolean value by applying the Boolean function of this gate to the values of the incoming edges. The Boolean value computed by the output gate is denoted by $f_C(a_1, \ldots, a_{\alpha(C)})$. In such a way the B-circuit C computes the $\alpha(C)$-ary Boolean function f_C. For more formal definitions see [Vol99]. Furthermore, let $\mathrm{BF}[B] =_{\mathrm{df}} \{f_C \mid C \text{ is a } B\text{-circuit}\}$ be the class of all Boolean functions which can be computed by B-circuits.

Now let $\mathbf{A}$ be a property related to Boolean functions. For a finite set B of

Boolean functions define

$$\mathbf{A}(B) =_{\text{df}} \{(C,a) \mid C \text{ is a } B\text{-circuit such that } (f_C, a) \text{ has property } \mathbf{A}\}.$$

In order to study the complexity of $\mathbf{A}(B)$ the following question is very important: How can we relate the complexity of $\mathbf{A}(B)$ and $\mathbf{A}(B')$ by relating the sets B and B' of Boolean functions themselves? The following proposition gives a satisfactory answer.

Proposition 1 *Let* $\mathbf{A}$ *be a property of Boolean functions. For finite sets* B *and* B' *of Boolean functions, if* $B \subseteq \mathrm{BF}[B']$ *then* $\mathbf{A}(B) \leq_{\mathrm{m}}^{\log} \mathbf{A}(B')$. *Consequently, if* $\mathrm{BF}[B] = \mathrm{BF}[B']$ *then* $\mathbf{A}(B) \equiv_{\mathrm{m}}^{\log} \mathbf{A}(B')$.

Proof: To convert a B-circuit into an equivalent B'-circuit just replace every B-gate by a B'-circuit computing it. □

In what follows we will make substantial use of the fact that the complexity of $\mathbf{A}(B)$ depends only on $\mathrm{BF}[B]$. To this end it is necessary to study the classes of Boolean functions which have the form $\mathrm{BF}[B]$. It turns out that this are exactly those classes of Boolean functions which contain the identity function and which are closed under superposition (i.e., substitution, permutation of variables, identification of variables, and introduction of non-essential variables). In the following section we will report on important results which are known on these classes.

3 Closed Classes of Boolean Functions

A function $f : \{0,1\}^n \to \{0,1\}$ with $n \geq 0$ is called an *n-ary Boolean function.* By BF we denote the class of all Boolean functions. In particular, let c_0 and c_1 be the 0-ary constant functions having value 0 and 1, resp., let id and non be the unary functions defined by $\mathrm{id}(a) = a$ and $\mathrm{non}(a) = 1 \Leftrightarrow a = 0$, let et, vel, and aut be the binary functions defined by $\mathrm{et}(a,b) = 1 \Leftrightarrow a = b = 1$, $\mathrm{vel}(a,b) = 0 \Leftrightarrow a = b = 0$, and $\mathrm{aut}(a,b) = 1 \Leftrightarrow a \neq b$. We also write 0 instaed of c_0, 1 instaed of c_1, $\overline{x}$ or $\neg x$ instead of $\mathrm{non}(x)$, $x \wedge y$, $x \cdot y$, or xy instead of $\mathrm{et}(x,y)$, $x \vee y$ instead of $\mathrm{vel}(x,y)$, and $x \oplus y$ instead of $\mathrm{aut}(x,y)$. For $i \in \{1,2,\ldots,n\}$, the i-th variable of the n-ary Boolean function f is said to be *non-essential* iff $f(a_1,\ldots a_{i-1},0,a_{i+1},\ldots,a_n) = f(a_1,\ldots a_{i-1},1,a_{i+1},\ldots,a_n)$ for all $a_1,\ldots a_{i-1},a_{i+1},\ldots,a_n \in \{0,1\}$.

For a set B of Boolean functions let $[B]$ be the smallest class which contains $B \cup \{\mathrm{id}\}$ and which is closed under superposition (i.e. substitution, permutation of variables and identification of variables, introduction of non-essential variables). A set B of Boolean functions is called a *base* of the class F of

Boolean functions if $[B] = F$ and it is called *complete* if $[B] = \mathrm{BF}$. A class F of Boolean functions is called *closed* if $[F] = F$.

The closed classes of Boolean functions are closely related to the sets of Boolean functions computed by B-circuits, as shown in the following "folklore" statement.

Proposition 2 *For any set B of Boolean functions,* $\mathrm{BF}[B] = [B]$.

Now consider some special properties of Boolean functions. An n-ary Boolean function f is said to be

- *a-reproducing* iff $f(a, a, \ldots, a) = a$ $(a \in \{0, 1\})$,
- *linear* iff there exist $a_0, a_1, \ldots, a_n \in \{0, 1\}$ such that $f(b_1, \ldots, b_n) = a_0 \oplus (a_1 \cdot b_1) \oplus (a_2 \cdot b_2) \oplus \ldots \oplus (a_n \cdot b_n)$ for all $b_0, b_1, \ldots, b_n \in \{0, 1\}$,
- *self-dual* iff $f(a_1, \ldots, a_n) = \overline{f(\overline{a_1}, \ldots \overline{a_n})}$ for all $a_1, \ldots, a_n \in \{0, 1\}$,
- *monotone* iff $f_m(a_1, \ldots, a_n) \leq f_m(b_1, \ldots, b_n)$ for all $a_1, \ldots, a_n, b_1, \ldots, b_n \in \{0, 1\}$ such that $a_1 \leq b_1, a_2 \leq b_2, \ldots, a_n \leq b_n$,
- *a-separating* iff there exists an $i \in \{1, 2, \ldots, n\}$ such that $f^{-1}(a) \subseteq \{0, 1\}^{i-1} \times \{a\} \times \{0, 1\}^{n-i}$ $(a \in \{0, 1\})$,
- *a-separating of degree m* iff for every $U \subseteq f^{-1}(a)$ such that $|U| = m$ there exists an $i \in \{1, 2, \ldots, n\}$ such that $U \subseteq \{0, 1\}^{i-1} \times \{a\} \times \{0, 1\}^{n-i}$ $(a \in \{0, 1\}, m \geq 2)$.

The classes of all Boolean functions which are 0-reproducing, 1-reproducing, linear, self-dual, monotone, 0-separating, 1-separating, 0-separating of degree m, and 1-separating of degree m, resp., are denoted by $\mathrm{BF}, \mathrm{R}_0, \mathrm{R}_1, \mathrm{L}, \mathrm{D}, \mathrm{M}, \mathrm{S}_0, \mathrm{S}_1, \mathrm{S}_0^m$, and S_1^m, resp.

The closed classes of Boolean functions were intensively studied by E. L. Post already at the beginning of the twenties. He gave a complete characterization of these classes. In this paper we will make substantial use of his main results which are presented in the following two theorems. For a detailed presentation see also [JGK70], the non-German reader may prefer [Pip97].

Theorem 3 ([Pos41]) *1. The complete list of closed classes of Boolean functions is:*
$\mathrm{BF}, \mathrm{R}_0, \mathrm{R}_1, \mathrm{R} =_{\mathrm{df}} \mathrm{R}_0 \cap \mathrm{R}_1$,
$\mathrm{M}, \mathrm{M}_0 =_{\mathrm{df}} \mathrm{M} \cap \mathrm{R}_0, \mathrm{M}_1 =_{\mathrm{df}} \mathrm{M} \cap \mathrm{R}_1, \mathrm{M}_2 =_{\mathrm{df}} \mathrm{M} \cap \mathrm{R}$,
$\mathrm{D}, \mathrm{D}_1 =_{\mathrm{df}} \mathrm{D} \cap \mathrm{R}, \mathrm{D}_2 =_{\mathrm{df}} \mathrm{D} \cap \mathrm{M}$,
$\mathrm{L}, \mathrm{L}_0 =_{\mathrm{df}} \mathrm{L} \cap \mathrm{R}_0, \mathrm{L}_1 =_{\mathrm{df}} \mathrm{L} \cap \mathrm{R}_1, \mathrm{L}_2 =_{\mathrm{df}} \mathrm{L} \cap \mathrm{R}, \mathrm{L}_3 =_{\mathrm{df}} \mathrm{L} \cap \mathrm{D}$,

S_0, $S_{02} =_{df} S_0 \cap R$, $S_{01} =_{df} S_0 \cap M$, $S_{00} =_{df} S_0 \cap R \cap M$,
S_1, $S_{12} =_{df} S_1 \cap R$, $S_{11} =_{df} S_1 \cap M$, $S_{10} =_{df} S_1 \cap R \cap M$,
S_0^m, $S_{02}^m =_{df} S_0^m \cap R$, $S_{01}^m =_{df} S_0^m \cap M$, $S_{00}^m =_{df} S_0^m \cap R \cap M$ *for* $m \geq 2$,
S_1^m, $S_{12}^m =_{df} S_1^m \cap R$, $S_{11}^m =_{df} S_1^m \cap M$, $S_{10}^m =_{df} S_1^m \cap R \cap M$ *for* $m \geq 2$,
$E =_{df} [\mathrm{et}] \cup [c_0] \cup [c_1]$, $E_0 =_{df} [\mathrm{et}] \cup [c_0]$, $E_1 =_{df} [\mathrm{et}] \cup [c_1]$, $E_2 =_{df} [\mathrm{et}]$,
$V =_{df} [\mathrm{vel}] \cup [c_0] \cup [c_1]$, $V_0 =_{df} [\mathrm{vel}] \cup [c_0]$, $V_1 =_{df} [\mathrm{vel}] \cup [c_1]$, $V_2 =_{df} [\mathrm{vel}]$,
$N =_{df} [\mathrm{non}] \cup [c_0] \cup [c_1]$, $N_2 =_{df} [\mathrm{non}]$,
$I =_{df} [\mathrm{id}] \cup [c_0] \cup [c_1]$, $I_0 =_{df} [\mathrm{id}] \cup [c_0]$, $I_1 =_{df} [\mathrm{id}] \cup [c_1]$, $I_2 =_{df} [\mathrm{id}]$,
$C =_{df} [c_0] \cup [c_1]$, $C_0 =_{df} [c_0]$, $C_1 =_{df} [c_1]$, $\emptyset$.

2. *The inclusional relationships between the closed classes of Boolean functions are presented in Figure 1.*

3. *There exists an algorithm which, given a finite set $B \subseteq \mathrm{BF}$, determines the closed class of Boolean functions from the list above which coincides with $[B]$.*

4. *There exists an algorithm which, given $f \in \mathrm{BF}$ and a finite set $B \subseteq \mathrm{BF}$, decides whether $f \in [B]$ or not.*

5. *Every closed class of Boolean functions has a finite base.*

Let us consider an example. Define the Boolean function f^3 such that $f(x, y, z) = 1$ iff exactly one argument has the value 1. Clearly, f is 0-reproducing but not 1-reproducing. Now assume that f is a linear function, i.e., $f(x, y, z) = a_0 \oplus (a_1 \cdot x) \oplus (a_2 \cdot y) \oplus (a_3 \cdot z)$ for suitable $a_0, a_1, a_2, a_3 \in \{0, 1\}$. Since $f(0, 0, 0) = 0$ we know $a_0 = 0$ and it follows that $a_1 = a_2 = a_3 = 1$, because of $f(1, 0, 0) = f(0, 1, 0) = f(0, 0, 1) = 1$. But this is a contradiction because $f(1, 1, 1) = 0$, showing that f is not linear. Furthermore f is not monotone since $f(1, 0, 0) = 1$ and $f(1, 1, 1) = 0$, and it is not self-dual because of $f(0, 0, 0) = f(1, 1, 1)$. Finally, f cannot be 1-separating of degree 2 because of $f(0, 0, 1) = f(0, 1, 0) = 1$. Summarizing the above, f is not in R_1, L, M, D, and S_1^2 but it is in R_0. A short look at Figure 1 shows $[\{f\}] = R_0$.

For an n-ary Boolean function f define the Boolean function dual(f) by $\mathrm{dual}(f)(x_1, \ldots, x_n) =_{df} \overline{f(\overline{x_1}, \ldots, \overline{x_n})}$. The functions f and dual(f) are said to be *dual*. Obviously, $\mathrm{dual}(\mathrm{dual}(f)) = f$. Furthermore, f is self-dual iff $\mathrm{dual}(f) = f$. For a class F of Boolean functions define $\mathrm{dual}(F) =_{df} \{\mathrm{dual}(f) \mid f \in F\}$. The classes F and dual(F) are called *dual*.

Proposition 4 1. *If B is a set of Boolean functions then* $[\mathrm{dual}(B)] = \mathrm{dual}([B])$.

2. *Every closed class is dual to its "mirror class" (via the symmetry axis in Figure 1).*

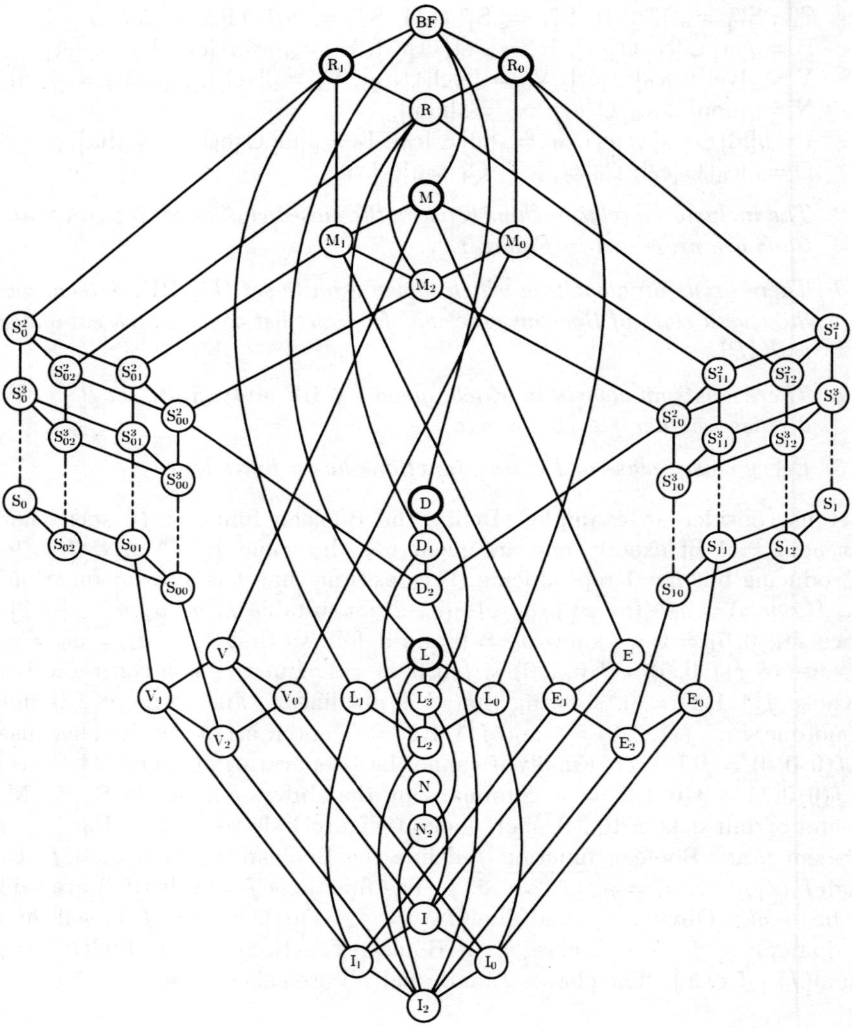

Figure 1: Graph of all classes of Boolean function being closed under superposition

4 Circuit Value

Let B be a finite set of Boolean functions. The *circuit value problem for B-circuits* is defined as

$$\mathbf{VAL}(B) =_{\text{df}} \{(C,a) \mid C \text{ is a } B\text{-circuit}, a \in \{0,1\}^{\alpha(C)}, \text{ and } f_C(a) = 1\}$$

It is obvious that $\mathbf{VAL}(B) \in \text{P}$ for every finite set B of Boolean functions. LADNER [Lad77] proved that $\mathbf{VAL}(\{\text{et}, \text{vel}, \text{non}\})$ is $\leq_{\text{m}}^{\log}$-complete for P. GOLDSCHLAGER [Gol77] proved that even $\mathbf{VAL}(\{\text{et}, \text{vel})$ is $\leq_{\text{m}}^{\log}$-complete for P. Eventually, GOLDSCHLAGER and PARBERRY proved for any set B of binary Boolean functions: If $\{\text{et}, \text{vel}\} \subseteq [\text{B}]$ or ($B \not\subseteq \text{L}$ and $B \not\subseteq \text{M}$) then $\mathbf{VAL}(B)$ is $\leq_{\text{m}}^{\log}$-complete for P, otherwise $\mathbf{VAL}(B)$ is acceptable in $\log^2 n$ space. This is already a "general" result but it applies only to binary Boolean functions and it disdinguishes only between "P-complete" and "acceptable in $\log^2 n$ space". To obtain a more general and more refined result we strengthen Proposition 1 for the case of the circuit value problem.

Proposition 5 *For finite sets B and B' of Boolean functions, if $B \subseteq [B' \cup \{\text{id}, \text{c}_0, \text{c}_1\}]$ then $\mathbf{VAL}(B) \leq_{\text{m}}^{\log} \mathbf{VAL}(B')$. Consequently, if $[B \cup \{\text{id}, \text{c}_0, \text{c}_1\}] = [\text{B}' \cup \{\text{id}, \text{c}_0, \text{c}_1\}]$ then $\mathbf{VAL}(B) \equiv_{\text{m}}^{\log} \mathbf{VAL}(B')$.*

Proof: As for Proposition 1, but additionally a c_0-gate (c_1-gate, resp.) is replaced by an input gate with Boolean value 0 (1, resp.). □

Hence, for the study of the complexity of $\mathbf{VAL}(B)$, only those closed classes of Boolean functions are of importance which contain c_0, and c_1. A close inspection of Figure 1 shows:

Proposition 6 *The closed classes of Boolean functions containing c_0, and c_1 are* BF, M, V, E, L, N, *and* I.

Now we are ready to prove our main theorem on the complexity of $\mathbf{VAL}(B)$.

Theorem 7 *Let B be a finite set of Boolean functions.*

if *($B \subseteq \text{N}$)* ***then***
$\mathbf{VAL}(B) \in \text{L}$
else if *(($B \subseteq \text{E}$)* ***or*** *($B \subseteq \text{V}$))* ***then***
$\mathbf{VAL}(B)$ *is $\leq_{\text{m}}^{\log}$-complete for* NL
else if *($B \subseteq \text{L}$)* ***then***
$\mathbf{VAL}(B)$ *is $\leq_{\text{m}}^{\log}$-complete for* $\oplus$L
else
$\mathbf{VAL}(B)$ *is $\leq_{\text{m}}^{\log}$-complete for* P

There exists an algorithm which decides which of the cases above holds.

Proof sketch: The proof runs along the following lines:

1. If $B \subseteq \mathrm{N}$ then $\mathbf{VAL}(B) \leq_{\mathrm{m}}^{\log} \mathbf{VAL}(\{\mathrm{non}\})$. We prove that $\mathbf{VAL}(\{\mathrm{non}\})$ is in L.

2. Let $B \not\subseteq \mathrm{N}$ and $B \subseteq \mathrm{V}$. We prove that $\mathbf{VAL}(B) \equiv_{\mathrm{m}}^{\log} \mathbf{VAL}(\{\mathrm{vel}\})$ and that $\mathbf{VAL}(\{\mathrm{vel}\})$ is $\leq_{\mathrm{m}}^{\log}$-equivalent to the NL-complete *graph accessibility problem.*

3. The case $B \not\subseteq \mathrm{N}$ and $B \subseteq \mathrm{E}$ is treated in the same way.

4. Let $B \not\subseteq \mathrm{N}$ and $B \subseteq \mathrm{L}$. We prove that $\mathbf{VAL}(B) \equiv_{\mathrm{m}}^{\log} \mathbf{VAL}(\{\mathrm{aut}\})$ and that $\mathbf{VAL}(\{\mathrm{aut}\})$ is $\leq_{\mathrm{m}}^{\log}$-equivalent to the $\oplus$L-complete *graph odd accessibility problem* (e.g., the problem of whether the number of paths from the start node to the target node in a given directed graph is odd).

5. Let $B \not\subseteq \mathrm{V}$, $B \not\subseteq \mathrm{E}$, and $B \not\subseteq \mathrm{L}$. We prove $\mathbf{VAL}(\{\mathrm{et}, \mathrm{vel}\}) \leq_{\mathrm{m}}^{\log} \mathbf{VAL}(B)$. Then $\mathbf{VAL}(B)$ is $\leq_{\mathrm{m}}^{\log}$-complete for P by the GOLDSCHLAGER result. □

Finally let us mention that the GOLDSCHLAGER-PARBERRY criterion for sets B of binary Boolean functions is not valid for arbitrary finite sets B of Boolean functions. Take for example $B = \{xy \vee xz\}$ for which we obtain vel $\notin$ [B] and $B \subseteq \mathrm{M}$. The GOLDSCHLAGER-PARBERRY criterion would yield that $\mathbf{VAL}(B)$ is acceptable in $\log^2 n$ space. However, Theorem 7 shows that $\mathbf{VAL}(B)$ is $\leq_{\mathrm{m}}^{\log}$-complete for P which is not likely to hold at the same time.

5 Satisfiability and Tautology

In this section we study the complexity of the satisfiability problem and the tautology problem for B-circuits. For a finite set B of Boolean functions we define

$$\mathbf{SAT}(B) =_{\mathrm{df}} \{C \mid C \text{ is a } B\text{-circuit and } f_C(a) = 1 \text{ for at least one } a \in \{0,1\}^{\alpha(C)}\}$$

and

$$\mathbf{TAUT}(B) =_{\mathrm{df}} \{C \mid C \text{ is a } B\text{-circuit and } f_C(a) = 1 \text{ for all } a \in \{0,1\}^{\alpha(C)}\}$$

It is obvious that $\mathbf{SAT}(B) \in \mathrm{NP}$ and $\mathbf{TAUT}(B) \in \text{co-NP}$ for every finite set B of Boolean functions. COOK [Coo71] proved that $\mathbf{SAT}(\{\mathrm{et}, \mathrm{vel}, \mathrm{non}\})$ is $\leq_{\mathrm{m}}^{\log}$-complete for NP. Consequently, $\mathbf{TAUT}(\{\mathrm{et}, \mathrm{vel}, \mathrm{non}\})$ is $\leq_{\mathrm{m}}^{\log}$-complete for co-NP. LEWIS [Lew79] proved that $\mathbf{SAT}(B)$ is $\leq_{\mathrm{m}}^{\log}$-complete for NP for every B such that $x \wedge \overline{y} \in [B]$.

The following easy-to-get reductions between different satisfiability and value problems are needed for the proof of the main theorem on the complexity of $\mathbf{SAT}(B)$.

Proposition 8 *Let B be a finite set of Boolean functions.*

1. *If $B \subseteq \mathrm{M}$ then* $\mathbf{SAT}(B) \leq_{\mathrm{m}}^{\log} \mathbf{VAL}(B)$.
2. *If $\mathrm{c}_0 \in [B] \subseteq \mathrm{M}$ then* $\mathbf{VAL}(B) \leq_{\mathrm{m}}^{\log} \mathbf{SAT}(B)$.
3. *If $\mathrm{c}_0, \mathrm{c}_1 \in [B]$ then* $\mathbf{VAL}(B) \leq_{\mathrm{m}}^{\log} \mathbf{SAT}(B)$.
4. $\mathbf{SAT}(B \cup \{\mathrm{c}_1\}) \leq_{\mathrm{m}}^{\log} \mathbf{SAT}(B)$.

Now we are ready to prove the main theorem on the complexity of $\mathbf{SAT}(B)$.

Theorem 9 *Let be B a finite set of Boolean functions.*

if *$((B \subseteq \mathrm{R}_1)$* ***or*** *$(B \subseteq \mathrm{D})$* ***or*** *$(B \subseteq \mathrm{N}))$* ***then***
  $\mathbf{SAT}(B) \in \mathrm{L}$
else if *$((B \subseteq \mathrm{E})$* ***or*** *$(B \subseteq \mathrm{V}))$* ***then***
    $\mathbf{SAT}(B)$ *is $\leq_{\mathrm{m}}^{\log}$-complete for* NL
else if *$(B \subseteq \mathrm{L})$* ***then***
    $\mathbf{SAT}(B)$ *is $\leq_{\mathrm{m}}^{\log}$-complete for* $\oplus$L
else if *$(B \subseteq \mathrm{M})$* ***then***
    $\mathbf{SAT}(B)$ *is $\leq_{\mathrm{m}}^{\log}$-complete for* P
else
   $\mathbf{SAT}(B)$ *is $\leq_{\mathrm{m}}^{\log}$-complete for* NP.

There exists an algorithm which decides which of the cases above holds.

Proof sketch: The proof runs along the following lines:
1. If $B \subseteq \mathrm{R}_1$ then every B-circuit C is satisfiable because of $f_C(1, \ldots 1) = 1$.
2. If $B \subseteq D$ then every B-circuit C is satisfiable because of $f_C(0, \ldots, 0) = 1$ or $f_C(1, \ldots, 1) = 1$.
3. Let $B \subseteq \mathrm{N}$. We prove that $\mathbf{SAT}(B) \leq_{\mathrm{m}}^{\log} \mathbf{SAT}(\{\mathrm{non}, \mathrm{c}_1\})$, and that $\mathbf{SAT}(\{\mathrm{non}, \mathrm{c}_1\})$ is in L.
4. Let $B \not\subseteq \mathrm{R}_1$, $B \not\subseteq \mathrm{N}$, and $B \subseteq \mathrm{V}$. We prove $\mathbf{SAT}(B) \equiv_{\mathrm{m}}^{\log} \mathbf{VAL}(B)$. Now Theorem 7 yields that $\mathbf{SAT}(B)$ is $\leq_{\mathrm{m}}^{\log}$-complete for NL.
5. The case $B \not\subseteq \mathrm{R}_1$, $B \not\subseteq \mathrm{N}$, and $B \subseteq \mathrm{E}$ is treated in the same way.
6. Let $B \not\subseteq \mathrm{R}_1$, $B \not\subseteq \mathrm{N}$, and $B \subseteq \mathrm{L}$. We prove that $\mathbf{SAT}(\{\mathrm{aut}\}) \leq_{\mathrm{m}}^{\log}$ $\mathbf{SAT}(\mathrm{B}) \leq_{\mathrm{m}}^{\log} \mathbf{SAT}(\{\mathrm{aut}, \mathrm{c}_1\})$, that $\mathbf{SAT}(\{\mathrm{aut}, \mathrm{c}_1\})$ is in $\oplus$L and that the $\oplus$L-complete graph odd accessibility problem is $\leq_{\mathrm{m}}^{\log}$-reducible to $\mathbf{SAT}(\{\mathrm{aut}\})$.
7. Let $B \not\subseteq \mathrm{R}_1$, $B \not\subseteq \mathrm{D}$, $B \not\subseteq \mathrm{V}$, $B \not\subseteq \mathrm{E}$, and $B \subseteq \mathrm{M}$. We prove $\mathbf{SAT}(B) \equiv_{\mathrm{m}}^{\log}$ $\mathbf{VAL}(B)$. Now Theorem 7 yields that $\mathbf{SAT}(B)$ is $\leq_{\mathrm{m}}^{\log}$-complete for P.
8. Let $B \not\subseteq \mathrm{R}_1$, $B \not\subseteq \mathrm{D}$, $B \not\subseteq \mathrm{L}$, and $B \not\subseteq \mathrm{M}$. We prove $x \wedge \overline{y} \in [B]$. By the LEWIS result we get that $\mathbf{SAT}(B)$ is $\leq_{\mathrm{m}}^{\log}$-complete for NP. □

Now turn to the tautology problem. The following obvious duality principle allows us to conclude the main theorem on the complexity of $\mathbf{TAUT}(B)$ from the corresponding theorem on $\mathbf{SAT}(B)$.

Proposition 10 *For any finite set B of Boolean functions,* $\mathbf{TAUT}(B) \equiv_{\mathrm{m}}^{\log} \mathbf{SAT}(\mathrm{dual}(B))$.

Now we are ready to prove the main theorem on the complexity of $\mathbf{TAUT}(B)$.

Theorem 11 *Let be B a finite set of Boolean functions.*

if *$((B \subseteq \mathrm{R}_0)$* ***or*** *$(B \subseteq \mathrm{D})$* ***or*** *$(B \subseteq \mathrm{N}))$* ***then***
  $\mathbf{TAUT}(B) \in \mathrm{L}$
else if *$((B \subseteq \mathrm{E})$* ***or*** *$(B \subseteq \mathrm{V}))$* ***then***
    $\mathbf{TAUT}(B)$ *is $\leq_{\mathrm{m}}^{\log}$-complete for* NL
else if *$(B \subseteq \mathrm{L})$* ***then***
    $\mathbf{TAUT}(B)$ *is $\leq_{\mathrm{m}}^{\log}$-complete for* $\oplus$L
else if *$(B \subseteq \mathrm{M})$* ***then***
    $\mathbf{TAUT}(B)$ *is $\leq_{\mathrm{m}}^{\log}$-complete for* P
else
  $\mathbf{TAUT}(B)$ *is $\leq_{\mathrm{m}}^{\log}$-complete for* co-NP.

There exists an algorithm which decides which of the cases above holds.

6 Quantifiers

In this section we study the complexity of problems defined by B-circuits with quantified input variables. For $Q_1, Q_2, \ldots, Q_m \in \{\exists, \forall\}$ and $k \geq 1$, we call $Q_1 Q_2 \ldots Q_m$ a *Σ_k-string* (a *Π_k-string*) if $Q_1 = \exists$ ($Q_1 = \forall$, resp.) and there are at most $k-1$ alternations between $\exists$ and $\forall$ in $Q_1 Q_2 \ldots Q_m$. For a finite set B of Boolean functions define:

$$\mathbf{\Sigma}_k(B) =_{\mathrm{df}} \{Q_1 x_1 \ldots Q_m x_m C \mid C \text{ is a } B\text{-circuit with Boolean variables } x_1, \ldots, x_m,\ Q_1 \ldots Q_m \text{ is a } \Sigma_k\text{-string and } Q_1 x_1 \ldots Q_m x_m f_C(x_1, x_2, \ldots, x_m) = 1\}$$

$$\mathbf{\Pi}_k(B) =_{\mathrm{df}} \{Q_1 x_1 \ldots Q_m x_m C \mid C \text{ is a } B\text{-circuit with Boolean variables } x_1, \ldots, x_m,\ Q_1 \ldots Q_m \text{ is a } \Pi_k\text{-string and } Q_1 x_1 \ldots Q_m x_m f_C(x_1, \ldots, x_m) = 1\}$$

$$\mathbf{QBC}(B) =_{\mathrm{df}} \{Q_1 x_1 \ldots Q_m x_m C \mid C \text{ is a } B\text{-circuit with Boolean variables } x_1, \ldots, x_m,\ Q_1, \ldots, Q_m \in \{\exists, \forall\} \text{ and } Q_1 x_1 \ldots Q_m x_m f_C(x_1, \ldots, x_m) = 1\}$$

Notice that $\mathbf{\Sigma}_1(B) = \mathbf{SAT}(B)$ and $\mathbf{\Pi}_1(B) = \mathbf{TAUT}(B)$ have already been treated in the previous section. Here we concentrate on the case $k \geq 2$.

It is obvious that $\mathbf{\Sigma}_k(B) \in \Sigma_{\mathrm{k}}^{\mathrm{p}}$, $\mathbf{\Pi}_k(B) \in \Pi_{\mathrm{k}}^{\mathrm{p}}$, and $\mathbf{QBC}(B) \in$ PSPACE for every finite set B of Boolean functions. Stockmeyer and Meyer [SM73] proved that $\mathbf{\Sigma}_k(\{\text{et}, \text{vel}, \text{non}\})$ is $\leq_{\text{m}}^{\text{log}}$-complete for $\Sigma_{\text{k}}^{\text{P}}$, $\mathbf{\Pi}_k(\{\text{et}, \text{vel}, \text{non}\})$ is $\leq_{\text{m}}^{\text{log}}$-complete for $\Pi_{\text{k}}^{\text{P}}$, and $\mathbf{QBC}(\{\text{et}, \text{vel}, \text{non}\})$ is $\leq_{\text{m}}^{\text{log}}$-complete for PSPACE.

For sets B of monotone Boolean functions the complexity of $\mathbf{\Sigma}_k(B)$, $\mathbf{\Pi}_k(B)$ and $\mathbf{QBC}(B)$ can easily be related to the complexity of $\mathbf{VAL}(B)$.

Proposition 12 *If $B \subseteq \text{M}$ is a finite set of Boolean functions then $\mathbf{\Sigma}_k(B) \equiv_{\text{m}}^{\text{log}} \mathbf{\Pi}_k(B) \equiv_{\text{m}}^{\text{log}} \mathbf{QBC}(B) \equiv_{\text{m}}^{\text{log}} \mathbf{VAL}(B)$ for every $k \geq 2$.*

Theorem 13 *Let be B a finite set of Boolean functions, and let $k \geq 2$.*

if *$(B \subseteq \text{N})$* ***then***
$\mathbf{\Sigma}_k(B), \mathbf{\Pi}_k(B), \mathbf{QBC}(B) \in \text{L}$
else if *$((B \subseteq \text{E})$* ***or*** *$(B \subseteq \text{V}))$* ***then***
$\mathbf{\Sigma}_k(B), \mathbf{\Pi}_k(B)$, *and* $\mathbf{QBC}(B)$ *are* $\leq_{\text{m}}^{\text{log}}$ *- complete for* NL
else if *$(B \subseteq \text{L})$* ***then***
$\mathbf{\Sigma}_k(B), \mathbf{\Pi}_k(B)$, *and* $\mathbf{QBC}(B)$ *are* $\leq_{\text{m}}^{\text{log}}$*-complete for* $\oplus$L
else if *$(B \subseteq \text{M})$* ***then***
$\mathbf{\Sigma}_k(B), \mathbf{\Pi}_k(B)$, *and* $\mathbf{QBC}(B)$ *are* $\leq_{\text{m}}^{\text{log}}$*-complete for* P
else
$\mathbf{\Sigma}_k(B)$ *is* $\leq_{\text{m}}^{\text{log}}$*-complete for* $\Sigma_{\text{k}}^{\text{P}}$
$\mathbf{\Pi}_k(B)$ *is* $\leq_{\text{m}}^{\text{log}}$*-complete for* $\Pi_{\text{k}}^{\text{P}}$
$\mathbf{QBC}(B)$ *is* $\leq_{\text{m}}^{\text{log}}$*-complete for* PSPACE.

There exists an algorithm which decides which of the cases above holds.

Proof sketch: The proof for $\mathbf{\Sigma}_k(B)$ runs along the following lines, the other cases are analogous.

1. Let $B \subseteq \text{N}$. We prove that $\mathbf{\Sigma}_k(B) \leq_{\text{m}}^{\text{log}} \mathbf{\Sigma}_k(\{\text{non}, \text{c}_1\})$, and that $\mathbf{\Sigma}_k(\{\text{non}, \text{c}_1\})$ is in L.

2. Let $B \not\subseteq \text{N}$ and ($B \subseteq \text{V}$ or $B \subseteq \text{E}$). We prove that $\mathbf{\Sigma}_k(B) \equiv_{\text{m}}^{\text{log}} \mathbf{VAL}(B)$. By Theorem 7 we obtain that these problems are $\leq_{\text{m}}^{\text{log}}$-complete for NL.

3. Let $B \not\subseteq \text{N}$ and $B \subseteq \text{L}$. We prove that $\mathbf{\Sigma}_k(\{h\}) \leq_{\text{m}}^{\text{log}} \mathbf{\Sigma}_k(B) \leq_{\text{m}}^{\text{log}} \mathbf{\Sigma}_k(\{\text{aut}, \text{c}_1\})$ where $h(x, y, z) =_{\text{df}} x \oplus y \oplus z$, that $\mathbf{\Sigma}_k(\{\text{aut}, \text{c}_1\})$ is in $\oplus$L, and that the complement of the graph odd accessibility problem is $\leq_{\text{m}}^{\text{log}}$-reducible to $\mathbf{\Sigma}_2(\{h\})$ (notice that $\oplus$L is closed under complement).

4. Let $B \not\subseteq \text{V}$, $B \not\subseteq \text{E}$, and $B \subseteq \text{M}$. We prove that $\mathbf{\Sigma}_k(B) \equiv_{\text{m}}^{\text{log}} \mathbf{VAL}(B)$. By Theorem 7 we obtain that these problems are $\leq_{\text{m}}^{\text{log}}$-complete for P.

5. Let $B \not\subseteq \text{M}$ and $B \not\subseteq \text{L}$. We prove that $B \cup \{\text{c}_0, \text{c}_1\}$ is complete, that there exists an $h \in [B]$ such that $h(x, u, 0, 1) = x \vee u$ and $h(x, u, v, z) \leq u$ for all $(v, z) \neq (0, 1)$, and that $\mathbf{\Sigma}_k(B \cup \{\text{c}_0, \text{c}_1\}) \leq_{\text{m}}^{\text{log}} \mathbf{\Sigma}_k(B \cup \{h\})$ for such an h. □

7 Counting Functions

In this section we study the complexity of functions which count the number of satisfying inputs to a B-circuit. For a B-circuit C define $\#C =_{\mathrm{df}} \#\{a \in \{0,1\}^{\alpha(C)} \mid f_C(a) = 1\}$, and for any finite set B of Boolean functions define the *counting function* $\#(B)$ by $\#(B)(C) =_{\mathrm{df}} \#C$ where C is a B-circuit.

It is obvious that $\#(B) \in \#\mathrm{P}$ for every finite set B of Boolean functions. Simon [Sim75] proved that $\#(\{\mathrm{et}, \mathrm{vel}, \mathrm{non}\})$ is $\leq_{\mathrm{m}}^{\mathrm{log}}$-complete for $\#\mathrm{P}$.

To study the complexity of $\#(B)$ in the general case we will use the following propositions which are analogous to Proposition 1 and Proposition 8.4 and which are proved in the same way.

Proposition 14 *For B and B' be finite sets of Boolean functions, if $B \subseteq \mathrm{BF}[B']$ then $\#(B) \leq_{\mathrm{m}}^{\mathrm{log}} \#(B')$. Consequently, if $\mathrm{BF}[B'] = \mathrm{BF}[B']$ then $\#(B) \equiv_{\mathrm{m}}^{\mathrm{log}} \#(B')$.*

Proposition 15 *If B is a finite set of Boolean functions containing* et *then $\#(B \cup \{\mathrm{c}_1\}) \leq_{\mathrm{m}}^{\mathrm{log}} \#(B)$.*

We will use the following obvious duality principle.

Proposition 16 *For any circuit C there holds $\#\mathrm{dual}(C) = 2^{\alpha(C)} - \#C$.*

Next we prove a lemma about the representation of $\#\mathrm{P}$ functions by $\#(B)$ functions. We omit the proof of this lemma.

Lemma 17 *Let B be a finite set of Boolean functions.*

1. *If $[B] \supseteq \mathrm{S}_1$ then $\#(B)$ is $\leq_{\mathrm{m}}^{\mathrm{log}}$-complete for $\#\mathrm{P}$, i.e., for every function $f \in \#\mathrm{P}$ there exists a logspace computable function h generating B-circuits such that $f(x) = \#h(x)$.*
2. *If $[B] \supseteq \mathrm{S}_{10}$ or $[B] \supseteq \mathrm{S}_{00}$ then $\#(B)$ is $\leq_{1-\mathrm{T}}^{\mathrm{log}}$-complete for $\#\mathrm{P}$. More specifically, for every function $f \in \#\mathrm{P}$ there exist logspace computable functions h_1, h_2 generating B-circuits and logspace computable functions g_1, g_2 such that $f(x) = \#h_1(x) - g_1(x) = g_2(x) - \#h_2(x)$.*

Now we can prove our main result on the complexity of $\#(B)$.

Theorem 18 *Let be B a finite set of Boolean functions.*

if *$((B \subseteq \mathrm{N})$* ***or*** *$(B \subseteq \mathrm{D}))$* ***then***
$\#(B) \in \mathrm{FL}$
else if *$((B \subseteq \mathrm{E})$* ***or*** *$(B \subseteq \mathrm{V}))$* ***then***
$\#(B)$ *is* $\leq^{\log}_{1-\mathrm{T}}$ *-complete for* $\mathrm{FL}^{\mathrm{NL}}_{||}[O(\log n)]$
else if *$(B \subseteq \mathrm{L})$* ***then***
$\#(B)$ *is* $\leq^{\log}_{1-\mathrm{T}}$ *-hard for* $\mathrm{FL}^{\oplus\mathrm{L}}_{||}[1]$ *and*
$\#(B) \in \mathrm{FL}^{\oplus\mathrm{L}}_{||}[2]$
else if *$(B \subseteq \mathrm{R}_1)$* ***then***
$\#(B)$ *is* $\leq^{\log}_{1-\mathrm{T}}$ *-complete for* $\#\mathrm{P}$,
but not $\leq^{\log}_{\mathrm{m}}$*-complete for* $\#\mathrm{P}$
else if *$(B \subseteq \mathrm{M})$* ***then***
$\#(B)$ *is* $\leq^{\log}_{1-\mathrm{T}}$ *-complete for* $\#\mathrm{P}$.
If $\#(B)$ *is* $\leq^{\log}_{\mathrm{m}}$*-complete for* $\#\mathrm{P}$ *then* $\mathrm{P} = \mathrm{NP}$.
else
$\#(B)$ *is* $\leq^{\log}_{\mathrm{m}}$*-complete for* $\#\mathrm{P}$.

There exists an algorithm which decides which of the cases above holds.

Proof sketch: The proof runs along the following line. 1. Let $B \subseteq \mathrm{N}$. We prove that $\#(B) \leq^{\log}_{\mathrm{m}} \#(\{\mathrm{non}, \mathrm{c}_1\})$ and that $\#(\{\mathrm{non}, \mathrm{c}_1\})$ can be computed in logspace.

2. If $B \subseteq \mathrm{D}$ and C is a B-circuit with n input variables then $\#C = 2^{n-1}$.

3. Let $B \not\subseteq \mathrm{N}$ and $B \subseteq \mathrm{E}$. We prove that $\#(\{\mathrm{et}\}) \leq^{\log}_{\mathrm{m}} \#(\mathrm{B}) \leq^{\log}_{\mathrm{m}} \#(\{\mathrm{et}, \mathrm{c}_0, \mathrm{c}_1\})$, that $\#(\{\mathrm{et}, \mathrm{c}_0, \mathrm{c}_1\})$ is in $\mathrm{FL}^{\mathrm{NL}}_{||}[O(\log n)]$, and that every function from $\mathrm{FL}^{\mathrm{NL}}_{||}[O(\log n)]$ is $\leq^{\log}_{\mathrm{m}}$-reducible to $\#(\{\mathrm{et}\})$.

4. The case $B \not\subseteq \mathrm{N}$ and $B \subseteq \mathrm{V}$ is done in an anologous manner.

5. Let $B \not\subseteq \mathrm{N}$, $B \not\subseteq \mathrm{D}$, and $B \subseteq \mathrm{L}$. We prove that $\#(\{\mathrm{aut}\}) \leq^{\log}_{\mathrm{m}} \#(\mathrm{B}) \leq^{\log}_{\mathrm{m}} \#(\{\mathrm{aut}, \mathrm{c}_1\})$, that $\#(\{\mathrm{aut}, \mathrm{c}_1\})$ is in $\mathrm{FL}^{\oplus\mathrm{L}}_{||}[2]$, and that $\#(\{\mathrm{aut}\})$ is $\leq^{\log}_{1-\mathrm{T}}$-hard for $\mathrm{FL}^{\oplus\mathrm{L}}_{||}[1]$.

6. Let $B \not\subseteq \mathrm{D}$, $B \not\subseteq \mathrm{V}$, $B \not\subseteq \mathrm{E}$, $B \not\subseteq \mathrm{L}$, and $B \subseteq \mathrm{R}_1$. An inspection of Figure 1 shows that $[B] \supseteq \mathrm{S}_{10}$ or $[B] \supseteq \mathrm{S}_{00}$. By Lemma 17 we obtain $\#(B)$ is $\leq^{\log}_{1-\mathrm{T}}$-complete for $\#\mathrm{P}$. Because of $B \subseteq \mathrm{R}_1$ we have $\#(B)(C) > 0$ for every B-circuit C. Hence $\#(B)$ cannot be $\leq^{\log}_{\mathrm{m}}$-complete for $\#\mathrm{P}$.

7. Let $B \not\subseteq \mathrm{D}$, $B \not\subseteq \mathrm{V}$, $B \not\subseteq \mathrm{E}$, $B \not\subseteq \mathrm{L}$, and $B \subseteq \mathrm{M}$. As above we obtain that $\#(B)$ is $\leq^{\log}_{1-\mathrm{T}}$-complete for $\#\mathrm{P}$. Assume that $\#(B)$ is $\leq^{\log}_{\mathrm{m}}$-complete for $\#\mathrm{P}$. For an arbitrary $A \in \mathrm{NP}$ there exists an $f \in \#\mathrm{P}$ such that $x \in A \Leftrightarrow f(x) > 0$ for every x. Since $\#(B)$ is $\leq^{\log}_{\mathrm{m}}$-complete for $\#\mathrm{P}$ there exists a logspace computable function h such that $f(x) = \#(B)(h(x))$

for every x. Consequently, $x \in A \Leftrightarrow \#(B)(h(x)) > 0 \Leftrightarrow h(x) \in \mathbf{SAT}(B)$. Because of $B \subseteq \mathrm{M}$ we obtain $A \in \mathrm{P}$ by Theorem 9.

8. Let $B \not\subseteq \mathrm{D}$, $B \not\subseteq \mathrm{L}$, $B \not\subseteq \mathrm{R}_1$, and $B \not\subseteq \mathrm{M}$. An inspection of Figure 1 shows that $[B] \supseteq \mathrm{S}_1$. By Lemma 17 we obtain that $\#(B)$ is $\leq_{\mathrm{m}}^{\log}$-complete for $\#\mathrm{P}$. □

8 The Threshold Problem

In this section we will study the complexity of the *threshold problem*

$$\mathbf{THR}(B) =_{\mathrm{df}} \{(C,k) \mid C \text{ is a } B\text{-circuit}, \ k \geq 0, \text{ and } \#C \geq k\}$$

for every finite set B of Boolean functions. Obviously, $\mathbf{THR}(B) \in \mathrm{PP}$. GILL [Gil77] proved that $\mathbf{THR}(\{\mathrm{et}, \mathrm{vel}, \mathrm{non}\})$ is $\leq_{\mathrm{m}}^{\log}$-complete for PP.

Now we are ready to prove the main theorem on the complexity of $\mathbf{THR}(B)$.

Theorem 19 *Let be B a finite set of Boolean functions.*

if** (($B \subseteq \mathrm{N}$) **or** ($B \subseteq \mathrm{D}$)) **then
  $\mathbf{THR}(B) \in \mathrm{L}$
else if** (($B \subseteq \mathrm{E}$) **or** ($B \subseteq \mathrm{V}$)) **then
  $\mathbf{THR}(B)$ *is* $\leq_{\mathrm{m}}^{\log}$*-complete for* NL
else if** ($B \subseteq \mathrm{L}$) **then
  $\mathbf{THR}(B)$ *is* $\leq_{\mathrm{m}}^{\log}$*-complete for* $\oplus$L
else
  $\mathbf{THR}(B)$ *is* $\leq_{\mathrm{m}}^{\log}$*-complete for* PP.

There exists an algorithm which decides which of the cases above holds.

Proof: 1. If $B \subseteq \mathrm{N}$ or $B \subseteq \mathrm{D}$ then, by Theorem 18, we have $\#(B) \in \mathrm{FL}$ and consequently $\mathbf{THR}(B) \in \mathrm{L}$.

2. Let $B \not\subseteq \mathrm{N}$ and $B \subseteq \mathrm{E}$. By Theorem 18 we obtain $\#(B) \in \mathrm{FL}_{||}^{\mathrm{NL}}$ and hence $\mathbf{THR}(B) \in \mathrm{L}^{\mathrm{NL}} = \mathrm{NL}$. For the lower bound we prove $\overline{\mathrm{GAP}} \leq_{\mathrm{m}}^{\log} \mathbf{THR}(\{\mathrm{et}\})$

3. The case $B \not\subseteq \mathrm{N}$ and $B \subseteq \mathrm{V}$ is done in an analogous manner.

4. The case $B \not\subseteq \mathrm{N}$, $B \not\subseteq \mathrm{D}$, and $B \subseteq \mathrm{L}$ is also done analogously.

5. Let $B \not\subseteq \mathrm{D}$, $B \not\subseteq \mathrm{V}$, $B \not\subseteq \mathrm{E}$, and $B \not\subseteq \mathrm{L}$. An inspection of Figure 1 shows that $[B] \supseteq \mathrm{S}_{10}$ or $[B] \supseteq \mathrm{S}_{00}$ For a set $A \in \mathrm{PP}$ there exists a $f \in \#\mathrm{P}$ such that $(x,k) \in A \Leftrightarrow f(x) \geq k$. By Lemma 17 there exist logspace computable functions g, h such that $f(x) = \#(B)(h(x)) - g(x)$. Consequently, $(x,k) \in A \Leftrightarrow \#(B)(h(x)) \geq g(x) + k \Leftrightarrow (h(x), g(x)+k) \in \mathbf{THR}(B)$. □

9 Conclusions

We have investigated the complexity of circuit-based combinatorial problems defined by circuits with gates from an arbitrary finite base of Boolean functions. What about the special case of Boolean formulae, i.e., "tree-like" Boolean circuits? Informally, all completeness results for complexity classes beyond P are the same as in the general case of circuits. The completeness results for complexity classes inside P do not remain valid in many cases because evaluating a formula is much easier than evaluating a circuit.

References

[CH96] N. Creignou and M. Hermann. Complexity of generalized satisfiability counting problems. *Information and Computation*, 125:1–12, 1996.

[CH97] N. Creignou and J.-J. Hebrard. On generating all solutions of generalized satisfiability problems, 1997.

[Coo71] S. A. Cook. The complexity if theorem proving procedures. In *Proc. 3rd Ann. ACM Symp. on Theory of Computation*, pages 151–158, 1971.

[Cre95] N. Creignou. A dichotomy theorem for maximum generalized satisfiability problems. *Journal of Computer and System Sciences*, 51:511–522, 1995.

[Gil77] J.T. Gill. Computational complexity of probabilistic turing machines. *SIAM Journal of Computation*, 6:675–695, 1977.

[Gol77] L. M. Goldschlager. The monotone and planar circuit value problems are log-space complete for P. *SIGACT News*, 9:25–29, 1977.

[GP86] L.M. Goldschlager and I. Parberry. On the construction of parallel computers from various bases of boolean functions. *Theoretical Computer Science*, 43:43–58, 1986.

[JGK70] S. W. Jablonski, G. P. Gawrilow, and W. B. Kudrajawzew. *Boolesche Funktionen und Postsche Klassen.* Akademie-Verlag, 1970.

[KST97] S. Khanna, M. Sudan, and L. Trevisan. Constraint satisfaction: The approximability of minimization problems. In *Proceedings 12th Computational Complexity Conference*, pages 282–296. IEEE Computer Society Press, 1997.

[Lad77] R. E. Ladner. The circuit value problem is logspace complete for P. *SIGACT News*, pages 18–20, 1977.

[Lew79] H. R. Lewis. Satisfiability Problems for Propositional Calculi. *Mathematical Systems Theory*, 13:45–53, 1979.

[Pip97] N. Pippenger. *Theories of Computability.* Cambridge University Press, Cambridge, 1997.

[Pos41] E. L. Post. The two-valued iterative systems of mathematical logic. *Annals of Mathematical Studies*, 5:1–122, 1941.

[RV00] S. Reith and H. Vollmer. Optimal Satisfiability for Propositional Calculi and Constraint Satisfaction Problems. In *Proc. 25th MFCS*, volume 1893 of *Lecture Notes in Computer Science*. Springer Verlag, 2000.

[Sch78] Th.J. Schaefer. The complexity of satisfiability problems. In *Proc. 10th ACM STOC*, pages 216–226. Association for Computing Machinery, 1978.

[Sim75] J. Simon. On some central problems in computational complexity. Doctoral Thesis, Dept. of Computer Science, Cornell University, 1975.

[SM73] L. J. Stockmeyer and A. R. Meyer. Word problems requiring exponential time. In *Proc. 35th Ann. ACM Symp. on Theory of Computation*, pages 1–9, 1973.

[Vol99] H. Vollmer. *Introduction to Circuit Complexity*. Springer, 1999.

The rational skimming theorem

Jacques Sakarovitch *

Abstract

We define the notion of $\mathbb{K}$-covering of automata with multiplicity (in a semiring $\mathbb{K}$) that extend the one of covering of automata. We make use of this notion, together with the Schützenberger construct that we have explained in a previous work and that we briefly recall here, in order to give a direct and constructive proof of a fundamental theorem on $\mathbb{N}$-rational power series.

In a previous work (*cf.* [4]), we have shown how a construction, proposed by Schützenberger (in [8] and [9]) in order to prove that rational functions are unambiguous, can be given a central position in the theory of relations and functions realized by finite automata. The other basic results such as the "Rational Cross-section Theorem", the "Rational Uniformisation Theorem" (that is dual to the preceeding one), and the "Decomposition Theorem " (of rational functions into sequential functions) appear then as direct and formal consequences of it.

We have explained that this construction is indeed *a construction on finite automata* and we have described it in the framework of *covering of automata* — which is derived from the notion of covering of graphs that was proposed by Stallings ([10]) — and which makes (in our opinion) the whole subject much clearer.

The purpose of the present communication is to extend the concept of covering to the one of $\mathbb{K}$-covering that apply to automata *with multiplicity* in a semiring $\mathbb{K}$. And to make use of this notion together with the Schützenberger construct quoted above, in order to establish another result, due to Schützenberger as well, and that we call the *Rational Skimming Theorem*.

*LTCI, UMR 5141 CNRS/ENST , Paris

Theorem 1 [7] *If s is a $\mathbb{N}$-rational power series on A^*, then the series s' obtained from s by substracting* 1 *to every non-zero coefficient of s, i.e. the series*

$$s' = s - \underline{\mathsf{supp}\, s}$$

is a $\mathbb{N}$-rational power series as well.

This result is not new, by far. In [2, Theorem VI.11.1], it is obtained as the consequence of the Rational Cross-section Theorem quoted above (and of some other results such as the division theorem). In [6, Theorem II.8.6] and in [1, Theorem V.2.1], more direct proofs are given (the attribution to Schützenberger is made in the latter reference).

The proof presented here is hopefully simpler than the preceeding ones and corresponds to an explicit construction on automata. A complete exposition of all that matter, $\mathbb{K}$-coverings and their use in the theory of $\mathbb{K}$-rational series will be found in [5].

1 The Schützenberger covering

We basically follow the definitions and notation of [2] which we use without further notice. Those that follow in this section and that are more original have been described in detail in [4].

A (finite) automaton over a finite alphabet A, $\mathcal{A} = \langle\, Q, A, E, I, T\,\rangle$, is a directed labelled graph where Q, I and T are respectively the (finite) sets of states, initial states and terminal states, and E is the set of labelled edges. The *language accepted* by $\mathcal{A}$, that is the set of the labels of the successful computations in $\mathcal{A}$, also called *the behaviour* of $\mathcal{A}$, is denoted by $|\mathcal{A}|$.

A *morphism* φ *from an automaton* $\mathcal{B} = \langle\, R, A, F, J, U\,\rangle$ *into an automaton* $\mathcal{A} = \langle\, Q, A, E, I, T\,\rangle$ is indeed a pair of mappings (both denoted by φ): one between the set of states $\varphi\colon R \to Q$, and one between the set of edges $\varphi\colon F \to E$, which are consistent with the structure of the automata, that is, for every f in F:

i) the origin of $f\varphi$ is the image (by φ) of the origin of f;

ii) the label of $f\varphi$ is equal to the label of f;

iii) and $\quad J\varphi \subseteq I \quad$ and $\quad U\varphi \subseteq T$.

These conditions imply that the image of a successful computation in $\mathcal{B}$ is a successful computation in $\mathcal{A}$, that their labels are equal, and thus that $|\mathcal{B}| \subseteq |\mathcal{A}|$ holds.

For every state q of an automaton $\mathcal{A} = \langle Q, A, E, I, T \rangle$, let us denote by $\mathsf{Out}_{\mathcal{A}}(q)$ the set[1] of edges of $\mathcal{A}$ *the origin of which* is q, that is edges that are "going out" of q. One defines dually $\mathsf{In}_{\mathcal{A}}(q)$ as the set of edges of $\mathcal{A}$ *the end of which* is q, that is edges that are "going in" q.

If φ is a morphism from $\mathcal{B} = \langle R, A, F, J, U \rangle$ into $\mathcal{A} = \langle Q, A, E, I, T \rangle$ then for every r in R, φ maps $\mathsf{Out}_{\mathcal{B}}(r)$ into $\mathsf{Out}_{\mathcal{A}}(r\varphi)$, and $\mathsf{In}_{\mathcal{B}}(r)$ into $\mathsf{In}_{\mathcal{A}}(r\varphi)$. We say that φ is *Out-surjective* (resp. *Out-bijective*, *Out-injective*) if for every r in R the restriction of φ to $\mathsf{Out}_{\mathcal{B}}(r)$ is surjective onto $\mathsf{Out}_{\mathcal{A}}(r\varphi)$ (resp. bijective between $\mathsf{Out}_{\mathcal{B}}(r)$ and $\mathsf{Out}_{\mathcal{A}}(r\varphi)$, injective). Accordingly, we say that φ is *In-surjective* (resp. *In-bijective*, *In-injective*) if for every r in R the restriction of φ to $\mathsf{In}_{\mathcal{B}}(r)$ is surjective onto $\mathsf{In}_{\mathcal{A}}(r\varphi)$ (resp. bijective between $\mathsf{In}_{\mathcal{B}}(r)$ and $\mathsf{In}_{\mathcal{A}}(r\varphi)$, injective).

Definition 1 *Let $\mathcal{B} = \langle R, A, F, J, U \rangle$ and $\mathcal{A} = \langle Q, A, E, I, T \rangle$; a morphism $\varphi\colon \mathcal{B} \to \mathcal{A}$ is a* covering *(resp. a* co-covering*) if the following conditions hold:*

i) *φ is* Out-bijective *(resp.* In-bijective*);*

ii) *for every i in I, there exists a* unique *j in J such that $j\varphi = i$ (resp. for every t in T, there exists a* unique *s in S such that $s\varphi = t$);*

iii) *$T\varphi^{-1} = U$ (resp. $I\varphi^{-1} = J$).*

Proposition 1 *Any covering (resp. any co-covering) $\varphi\colon \mathcal{B} \to \mathcal{A}$ induces a bijection between the successful computations in $\mathcal{B}$ and those in $\mathcal{A}$.* ∎

Theorem & Definition 2 *Let $\mathcal{A}$ be an automaton and $\mathcal{A}_{\mathsf{det}}$ the determinized automaton of $\mathcal{A}$. We call* Schützenberger covering of $\mathcal{A}$ *the accessible part $\mathcal{S}$ of $\mathcal{A}_{\mathsf{det}} \times \mathcal{A}$. Then:*

i) *$\pi_{\mathcal{A}}$ is a covering from $\mathcal{S}$ onto $\mathcal{A}$.*

ii) *$\pi_{\mathcal{A}_{\mathsf{det}}}$ is an In-surjective morphism from $\mathcal{S}$ onto $\mathcal{A}_{\mathsf{det}}$.* ∎

We call *immersion of $\mathcal{A}$* a sub-automaton of a covering of $\mathcal{A}$. From all these definitions and result, one derives easily the result which is the basis of the present work.

Corollary 2 *Let $\mathcal{A}$ be an automaton on A^*. Then there exists an* unambiguous *automaton that is equivalent to $\mathcal{A}$ and that is a sub-automaton of a covering of $\mathcal{A}$.*

---

[1] Stallings denotes it "$\mathrm{Star}_{\mathcal{A}}(q)$". As the star is the common denomination for the generated submonoid, we cannot keep it, though it nicely conveys the idea of "a set of edges going out" of q.

Proof. Let $\mathcal{S}$ be the Schützenberger covering of $\mathcal{A}$. As $\pi_{\mathcal{A}_{\mathsf{det}}}$ is *In-surjective* from $\mathcal{S}$ onto $\mathcal{A}_{\mathsf{det}}$, one can delete edges in $\mathcal{S}$ (and possibly suppress the quality of being terminal to some of its states) in such a way that the sub-automaton $\mathcal{T}$ that is obtained is a *co-covering of* $\mathcal{A}_{\mathsf{det}}$. The automaton $\mathcal{T}$ is then *unambiguous* — as there is a one-to-one correspondence between its successful computations and those of $\mathcal{A}_{\mathsf{det}}$ — and equivalent to $\mathcal{A}_{\mathsf{det}}$, hence to $\mathcal{A}$. ■

The essence of this statement lies of course in the fact that the quoted unambiguous automaton is at the same time *equivalent to* and an *immersion of* $\mathcal{A}$. For otherwise, the deterministic automaton $\mathcal{A}_{\mathsf{det}}$ associated to $\mathcal{A}$ by the subset construction is obviously unambiguous and equivalent to $\mathcal{A}$; but it can not be immersed in $\mathcal{A}$: there is no relationships between the pathes in $\mathcal{A}$ and those in $\mathcal{A}_{\mathsf{det}}$.

Example 1 : The Figure 1 represents an automaton $\mathcal{A}_1$ that accepts all words of $\{a,b\}^*$ which contain at least one b (vertically, on the left), its determinized automaton $\mathcal{A}_{1\mathsf{det}}$, the Schützenberger covering of $\mathcal{A}_1$, and the two possible immersions that can be derived from it. □

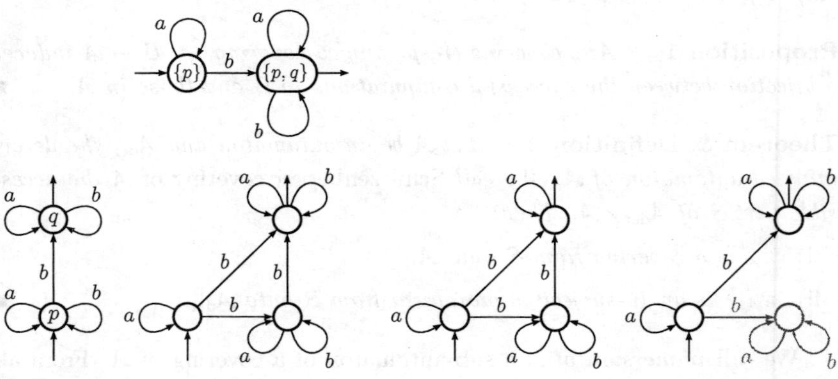

Figure 1: An automaton, its Schützenberger covering, and two immersions.

2 $\mathbb{K}$-automata

As far as *polynomials* and *power series* are concerned, we follow the definitions and notation of [1]. The set of polynomials over A^* with multiplicity in a semiring $\mathbb{K}$ is denoted by $\mathbb{K}\langle A^*\rangle$. A (finite) automaton $\mathcal{A}$ over A^* with multiplicity in a semiring $\mathbb{K}$, or $\mathbb{K}$-*automaton* for short, is a straightforward generalization of a classical automaton. It is adequatly described as a triple $\mathcal{A} = \langle\, I, E, T\,\rangle$ where

- E is a square matrix of dimension Q whose entries are polynomial over A^* with coefficients in $\mathbb{K}$, *i.e.* elements of $\mathbb{K}\langle A^*\rangle$.

- I and T are vectors of dimension Q (respectively a row vector and a column vector) with entries in $\mathbb{K}\langle A^*\rangle$.

The dimension Q is called *the set of states of* $\mathcal{A}$, every entry $E_{p,q}$ of E is the *label of the transition* that goes from p to q in $\mathcal{A}$.[2] The *behaviour* of $\mathcal{A}$, denoted by $|\mathcal{A}|$, is defined if and only if the star of the matrix E, E^*, is defined and it holds:

$$|\mathcal{A}| = I \cdot E^* \cdot T$$

A power series is $\mathbb{K}$-*rational* if and only if it is the behaviour of a finite $\mathbb{K}$-automaton[3].

A polynomial is *proper* if its constant term (*i.e.* the coefficient of 1_{A^*}) is zero, a $\mathbb{K}$-automaton $\mathcal{A} = \langle\, I, E, T\,\rangle$ is *proper* if every entry of E is proper and every entry of I and T are in $\mathbb{K}$. It is known that the behaviour of $\mathcal{A}$ is defined if and only if it is equivalent to a proper $\mathbb{K}$-automaton.

Example 2 : Let us consider the $\mathbb{N}$-automaton over $\{a,b\}^*$, $\mathcal{C}_1$, defined by:

$$\mathcal{C}_1 = \langle \begin{pmatrix} 1 & 0 \end{pmatrix}, \begin{pmatrix} a+b & b \\ 0 & 2a+2b \end{pmatrix}, \begin{pmatrix} 0 \\ 1 \end{pmatrix} \rangle$$

and represented (in two ways) at the Figure 2. If every word f of $\{a,b\}^*$ is viewed as the writing of an integer in the binary system, where a is interpreted as 0 and b as 1, then $\mathcal{C}_1$ "computes" the integer written by f, which we denote by $\overline{f}$, in the sense that

$$|\mathcal{C}_1| = \sum_{f \in A^*} \overline{f}\, f \qquad \text{and the first terms of } |\mathcal{C}_1| \text{ reads then}$$

$$|\mathcal{C}_1| = b + ab + 2\,ba + 3\,bb + aab + 2\,aba + 3\,abb + 4\,baa + 5\,bab + \cdots \quad \square$$

---

[2]This definition coincides then with the classical one when $\mathbb{K} = \mathbb{B}$, the Boolean semiring.

[3]In the context of this paper, we can take this statement as a definition for the $\mathbb{K}$-rational series.

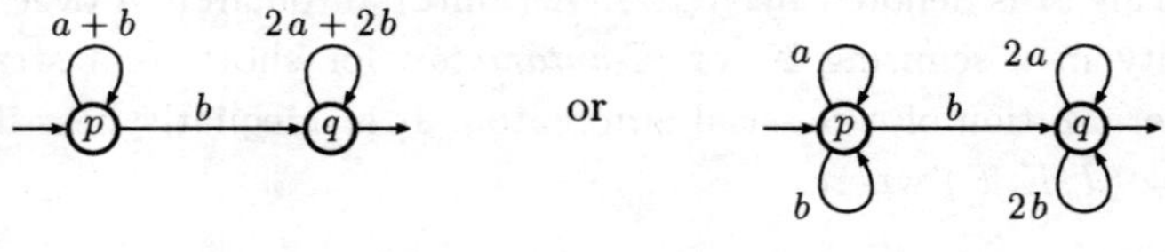

Figure 2: The $\mathbb{N}$-automaton $\mathcal{C}_1$.

The *support* of a power series, or of a polynomial, over A^* is the set of words of A^* whose coefficient is not zero in the series or in the polynomial. The *support* of a $\mathbb{K}$-automaton $\mathcal{A} = \langle I, E, T \rangle$ is the (classical) automaton obtained by taking the support of every entry of I, E and T. Conversely, to any (classical) automaton $\mathcal{A} = \langle Q, A, E, I, T \rangle$ is associated its *characteristic automaton* that is defined as the $\mathbb{K}$-automaton whose support is $\mathcal{A}$ and whose non-zero coefficients are all equal to $1_{\mathbb{K}}$ (generally, $\mathbb{K} = \mathbb{N}$).

Property 1 *The support of the behaviour of a $\mathbb{K}$-automaton $\mathcal{A}$ is contained in the behaviour of the support of $\mathcal{A}$. If $\mathbb{K}$ is a* positive *semiring, these two languages are equal.*

Property 2 *An automaton over A^* is unambiguous if and only if the behaviour of its characteristic $\mathbb{N}$-automaton is a characteristic power series.*

3 $\mathbb{K}$-coverings

The notion of covering seems to fit perfectly with the one of automaton with multiplicity. If $\mathcal{A}$ and $\mathcal{B}$ are two (classical) automata, the existence of a covering $\varphi\colon \mathcal{B} \to \mathcal{A}$ implies not only that $\mathcal{B}$ is equivalent to $\mathcal{A}$, *i.e.* that they both recognize the same language, but also that there exists a 1-to-1 correspondence between their successful computations, that is they are equivalent even if multiplicity is taken into account, *i.e.* they are equivalent as $\mathbb{N}$-automata — with the natural hypothesis that the label of every transition has multiplicity 1_N.

But it may be the case that we have two equivalent $\mathbb{K}$-automata $\mathcal{A}$ and $\mathcal{B}$ such that there exists an (automaton) morphism φ from the support of $\mathcal{B}$ into the support of $\mathcal{A}$ *which is not a covering.* As we said, an automaton morphism is a covering if its restriction to the corresponding "outgoing bouquets" is bijective. This condition is not adequate anymore for automata with multiplicity.

Example 3 : Let us consider the $\mathbb{N}$-automata $\mathcal{C}_2$ and $\mathcal{V}_2$ of the Figure 3. There exists an obvious morphism from the support of $\mathcal{C}_2$ onto the support of $\mathcal{V}_2$ that is not a covering: there is no bijection between $\mathsf{Out}_{\mathcal{C}_2}(j)$ and $\mathsf{Out}_{\mathcal{V}_2}(i)$, neither a co-covering: there is no bijection between $\mathsf{In}_{\mathcal{C}_2}(u)$ and $\mathsf{In}_{\mathcal{V}_2}(t)$.

These two $\mathbb{N}$-automata are equivalent nevertheless. The reason is that the *sum* of the labels of the transitions that go from j into the *set of states* whose image by φ is q is equal to the label of the transition that go from $i = j\varphi$ to q. □

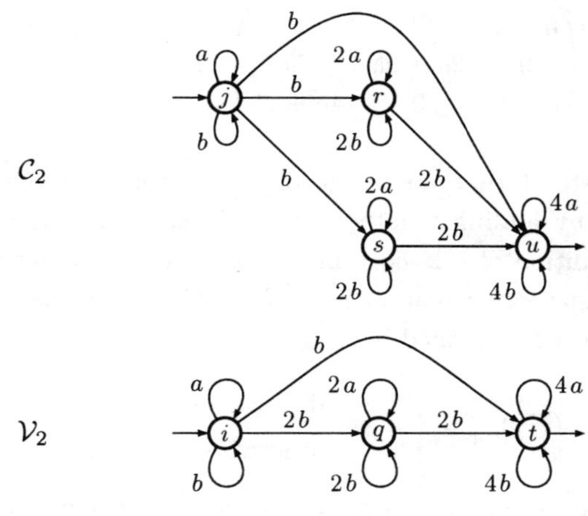

Figure 3: The *vertical* is a morphism, not a covering.

We now formalize the observation made in the example. Let $\mathcal{A} = \langle I, E, T \rangle$ and $\mathcal{B} = \langle J, F, U \rangle$ be two $\mathbb{K}$-automata of dimension Q and R respectively. Let $\varphi \colon R \to Q$ be a surjective mapping. Let F' be the matrix obtained from F by adding together the *columns* whose index have the same image by φ. Then φ is a $\mathbb{K}$-covering if any *row* of index r of F' is equal to the row of index $r\varphi$ of E.

Example 3 (continued) **:** If we write the above automata $\mathcal{C}_2$ and $\mathcal{V}_2$ as

$\mathcal{C}_2 = \langle J_2, F_2, U_2 \rangle$ and $\mathcal{V}_2 = \langle I_2, E_2, T_2 \rangle$, then:

$$F_2 = \begin{pmatrix} a+b & b & b & b \\ 0 & 2a+2b & 0 & 2b \\ 0 & 0 & 2a+2b & 2b \\ 0 & 0 & 0 & 4a+4b \end{pmatrix}.$$

We add the two mid columns and we get the matrix

$$F_2' = \begin{pmatrix} a+b & 2b & b \\ 0 & 2a+2b & 2b \\ 0 & 2a+2b & 2b \\ 0 & 0 & 4a+4b \end{pmatrix}$$

whose rows of index r and s are equal to the mid row of

$$E_2 = \begin{pmatrix} a+b & 2b & b \\ 0 & 2a+2b & 2b \\ 0 & 0 & 4a+4b \end{pmatrix}. \qquad \square$$

Once it is understood that an image of a $\mathbb{K}$-automaton under a $\mathbb{K}$-covering is obtained by adding together some of the "columns of the $\mathbb{K}$-automaton", the definition of a $\mathbb{K}$-covering is best written under a matrix expression. To any surjective mapping $\varphi\colon R \to Q$ we associate the *row monomial* $R \times Q$-matrix H_φ defined by:

$$(H_\varphi)_{r,q} = \begin{cases} 1 & \text{if } \; r\varphi = q \\ 0 & \text{otherwise} \end{cases}$$

Since φ is surjective, every column of H_φ contains at least one 1. From H_φ, a matrix K_φ is built by transposing H_φ and by making some entries equal to 0 in such a way that K_φ is *row monomial* (with exactly one 1 in every row). The matrix K_φ is not uniquely defined by φ (as is H_φ) but also by the arbitrary choice of a representative in every class modulo the mapping equivalence of φ.

Example 3 (continued) : If φ_2 is the mapping from $\{j, r, s, u\}$ onto $\{i, q, t\}$ such that $j\varphi_2 = i$, $u\varphi_2 = t$ and $r\varphi_2 = s\varphi_2 = q$, then:

$$H_{\varphi_2} = \begin{pmatrix} 1 & 0 & 0 \\ 0 & 1 & 0 \\ 0 & 1 & 0 \\ 0 & 0 & 1 \end{pmatrix} \quad \text{and} \quad K_{\varphi_2} = \begin{pmatrix} 1 & 0 & 0 & 0 \\ 0 & 1 & 0 & 0 \\ 0 & 0 & 0 & 1 \end{pmatrix} \quad \text{or} \quad \begin{pmatrix} 1 & 0 & 0 & 0 \\ 0 & 0 & 1 & 0 \\ 0 & 0 & 0 & 1 \end{pmatrix}$$

$\square$

The multiplication of an $X \times R$-matrix Z by H_φ *on the right* yields an $X \times Q$-matrix whose column q is the sum of the columns of Z of index s such that $s\varphi = q$. The multiplication of a $R \times X$-matrix Z by K_φ *on the left* yields a $Q \times X$-matrix whose row p is chosen among the rows of Z of index r such that $r\varphi = p$. We can then state:

Definition 2 *A mapping $\varphi\colon R \to Q$ is a $\mathbb{K}$-covering from $\mathcal{B} = \langle J, F, U \rangle$ onto $\mathcal{A} = \langle I, E, T \rangle$ if $\mathcal{A}$ is defined by:*

$$E = K_\varphi \cdot F \cdot H_\varphi\,, \qquad I = J \cdot H_\varphi \qquad \text{and} \qquad T = K_\varphi \cdot U \tag{1}$$

and if the following equations hold:

$$H_\varphi \cdot K_\varphi \cdot F \cdot H_\varphi = F \cdot H_\varphi \tag{2}$$

$$\text{and} \qquad H_\varphi \cdot K_\varphi \cdot U = U. \tag{3}$$

The definition of $\mathbb{K}$-covering *via* matrix expressions makes the proof of the following basic result particularly easy.

Proposition 3 *Any $\mathbb{K}$-automaton is equivalent to any of its $\mathbb{K}$-coverings.*

Proof. If $\varphi\colon \mathcal{B} \to \mathcal{A}$ is a $\mathbb{K}$-covering, it holds, for every n in $\mathbb{N}$:

$$\begin{aligned} I \cdot E^n \cdot T &= J \cdot H_\varphi \cdot (K_\varphi \cdot F \cdot H_\varphi)^n \cdot K_\varphi \cdot U && \text{by (1)} \\ &= J \cdot (H_\varphi \cdot K_\varphi \cdot F)^n \cdot H_\varphi \cdot K_\varphi \cdot U \\ &= J \cdot F^n \cdot H_\varphi \cdot K_\varphi \cdot U && \text{by (2)} \\ &= J \cdot F^n \cdot U && \text{by (3)} \end{aligned}$$

and this implies the equality $|\mathcal{B}| = J \cdot F^* \cdot U = I \cdot E^* \cdot T = |\mathcal{A}|$. ∎

To the $\mathbb{K}$-covering corresponds the *dual notion* of $\mathbb{K}$-co-covering. Roughly speaking, some rows will be added together, instead of some columns. More precisely we have:

Definition 3 *A mapping $\varphi\colon R \to Q$ is a $\mathbb{K}$-co-covering from $\mathcal{B} = \langle J, F, U \rangle$ onto $\mathcal{A} = \langle I, E, T \rangle$ if $\mathcal{A}$ is defined by:*

$$E = H^{\mathsf{t}}_\varphi \cdot F \cdot K^{\mathsf{t}}_\varphi\,, \qquad I = J \cdot K^{\mathsf{t}}_\varphi \qquad \text{and} \qquad T = H^{\mathsf{t}}_\varphi \cdot U$$

and if the following equations hold:

$$H^{\mathsf{t}}_\varphi \cdot F \cdot K^{\mathsf{t}}_\varphi \cdot H^{\mathsf{t}}_\varphi = H^{\mathsf{t}}_\varphi \cdot F \qquad \text{and} \qquad J \cdot K^{\mathsf{t}}_\varphi \cdot H^{\mathsf{t}}_\varphi = J.$$

Proposition 4
Any $\mathbb{K}$-automaton is equivalent to any of its $\mathbb{K}$-co-coverings. ■

Example 3 (continued) : Let us consider $\mathcal{C}_2 = \langle J_2, F_2, U_2 \rangle$ again: if we add the two mid rows of

$$F_2 = \begin{pmatrix} a+b & b & b & b \\ 0 & 2a+2b & 0 & 2b \\ 0 & 0 & 2a+2b & 2b \\ 0 & 0 & 0 & 4a+4b \end{pmatrix}$$

we get the matrix

$$\begin{pmatrix} a+b & b & b & b \\ 0 & 2a+2b & 2a+2b & 4b \\ 0 & 0 & 0 & 4a+4b \end{pmatrix}$$

whose columns r and s are equal to the mid column of the matrix

$$E_2' = \begin{pmatrix} a+b & b & b \\ 0 & 2a+2b & 4b \\ 0 & 0 & 4a+4b \end{pmatrix}$$

which defines *another* $\mathbb{N}$-automaton $\mathcal{V}_2' = \langle I_2, E_2', T_2 \rangle$ equivalent to $\mathcal{C}_2$ (*cf.* Figure 4). □

Coming back to our first intuition, we then have:

Property 3 *Let $\mathcal{A}$ and $\mathcal{B}$ be two (classical) automata and let $\varphi\colon \mathcal{B} \to \mathcal{A}$ be a covering (resp. a co-covering). Then, for any $\mathbb{K}$, φ is a $\mathbb{K}$-covering (resp. a $\mathbb{K}$-co-covering) from the characteristic automaton of $\mathcal{B}$ onto the characteristic automaton of $\mathcal{A}$.* ■

The following two properties are also easily verified.

Proposition 5 *Let $\mathcal{A}$ be a $\mathbb{K}$-automaton. Among all the $\mathbb{K}$-automata of which $\mathcal{A}$ is a $\mathbb{K}$-covering (resp. a $\mathbb{K}$-co-covering) there exists a unique one, effectively computable, that has a minimal number of states and of which all these $\mathbb{K}$-automata are $\mathbb{K}$-coverings (resp. $\mathbb{K}$-co-coverings).* ■

Proposition 6 *Let $\mathcal{A}$, $\mathcal{B}$ and $\mathcal{C}$ be three $\mathbb{K}$-automata such that $\mathcal{A}$ is a $\mathbb{K}$-covering of $\mathcal{C}$ and $\mathcal{B}$ is a $\mathbb{K}$-co-covering of $\mathcal{C}$. Then there exists a $\mathbb{K}$-automaton $\mathcal{D}$ which is a $\mathbb{K}$-co-covering of $\mathcal{A}$ and a $\mathbb{K}$-covering de $\mathcal{B}$.* ■

Remark 1 The terminology may be slightly misleading inasmuch as if a $\mathbb{K}$-automaton is exactly a classical automaton when $\mathbb{K} = \mathbb{B}$, a $\mathbb{B}$-covering *is not* a covering of classical automata, but only an Out-surjective morphism.

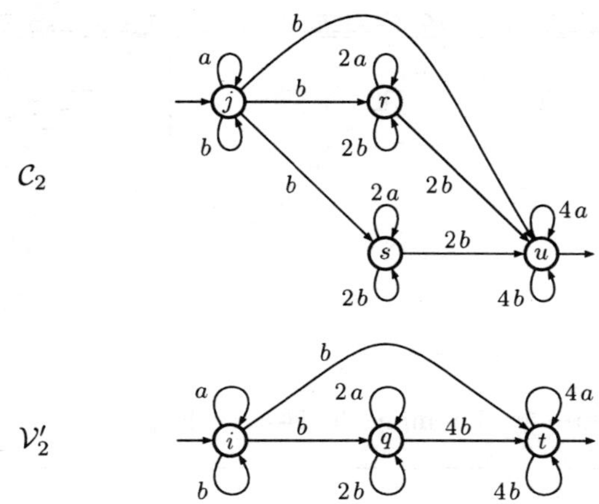

Figure 4: $\mathcal{C}_2$ is an $\mathbb{N}$-co-covering of $\mathcal{V}'_2$.

4 The skimming theorem

The Schützenberger construct, applied to a $\mathbb{N}$-automaton $\mathcal{A}$, yields an unambiguous $\mathbb{N}$-automaton $\mathcal{T}$ whose behaviour is the characteristic series of the support of the behaviour of $\mathcal{A}$. (*i.e.* $|\mathcal{T}| = \underline{\mathsf{supp}\,|\mathcal{A}|}$), and this is not surprising indeed. What is remarkable is that *the same construction*, together with the notion of $\mathbb{N}$-covering, yields a $\mathbb{N}$-automaton $\mathcal{P}$ which is the complement of $\mathcal{T}$ with respect to $\mathcal{A}$ *i.e.* $|\mathcal{A}| = |\mathcal{T}| + |\mathcal{P}|$, and this is the theorem we are aiming at:

Theorem 1 *If s is a $\mathbb{N}$-rational power series on A^*, the series* $s' = s - \underline{\mathsf{supp}\, s}$ *is a $\mathbb{N}$-rational power series as well.*

In other words, the series obtained by substracting 1 to every non zero coefficient of a $\mathbb{N}$-rational power series on A^* is still a $\mathbb{N}$-rational power series on A^* and this can be represented as on Figure 5. The series s is represented as the sequence of the values of the coefficients, adequatly oriented downwards; the upper layer is taken off[4]; what is left is the representation of *another* $\mathbb{N}$*-rational power series.*

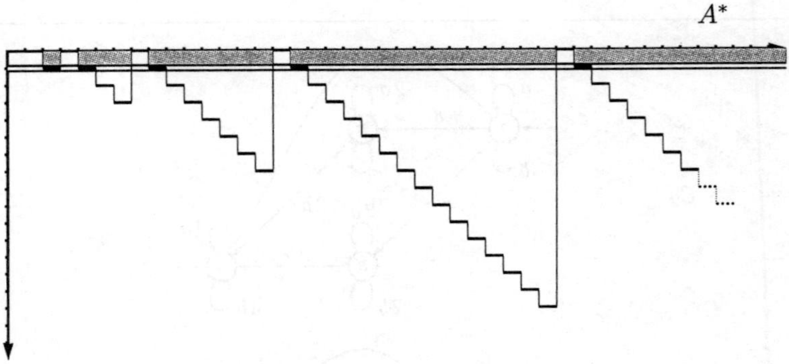

Figure 5: Skimming the $\mathbb{N}$-series $|\mathcal{C}_1|$.

Proof of Theorem 1. Let $\mathcal{A} = \langle\, I, E, T \,\rangle$ be a (proper) $\mathbb{N}$-automaton on A^* whose behaviour is equal to s, $\mathcal{S} = \langle\, J, F, U \,\rangle$ its Schützenberger covering (which is a $\mathbb{N}$-automaton of dimension R), and $\mathcal{T} = \langle\, J, G, V \,\rangle$ a S-immersion in $\mathcal{A}$, of dimension R as well. By definition, $\mathcal{T}$ is a sub-automaton of $\mathcal{S}$ and there exist a matrix H with coefficients in $\mathbb{N}\langle\langle A^* \rangle\rangle$ and a vector W with coefficients in $\mathbb{N}$ such that $F = G + H$ and $U = V + W$.

It is then observed that the automaton $\mathcal{S}'$ below, of dimension $R{\times}\{1,2,3\}$ is equivalent to $\mathcal{S}$, hence to $\mathcal{A}$:

$$\mathcal{S}' = \langle \begin{pmatrix} J & J & 0 \end{pmatrix}, \begin{pmatrix} G & 0 & H \\ 0 & G & 0 \\ 0 & 0 & F \end{pmatrix}, \begin{pmatrix} V \\ W \\ U \end{pmatrix} \rangle = \langle \begin{pmatrix} J & J & 0 \end{pmatrix}, F', \begin{pmatrix} V \\ W \\ U \end{pmatrix} \rangle \,.$$

Indeed, if we add *the rows* of $\mathcal{S}'$ of index $(r,1)$ and $(r,2)$ for every r in R we then get the matrices

$$\begin{pmatrix} J & J & 0 \end{pmatrix} \,, \quad \begin{pmatrix} G & G & H \\ 0 & 0 & F \end{pmatrix} \quad \text{and} \quad \begin{pmatrix} U \\ U \end{pmatrix} \,,$$

whose columns of index $(r,1)$ et $(r,2)$ (for every r in R) are equal: $\mathcal{S}'$ is a $\mathbb{N}$-*co-covering* of

$$\mathcal{S}'' = \langle \begin{pmatrix} J & 0 \end{pmatrix}, \begin{pmatrix} G & H \\ 0 & F \end{pmatrix}, \begin{pmatrix} U \\ U \end{pmatrix} \rangle \,.$$

[4] As one skims the cream from a milk jar.

The automaton $\mathcal{S}''$, of dimension $R\times\{1,2\}$, is itself an $\mathbb{N}$-*covering* of $\mathcal{S}$ since if we add *the columns* of index $(r,1)$ and $(r,2)$ (for every r in R) we get the matrices

$$J\ ,\quad \begin{pmatrix} F \\ F \end{pmatrix} \quad \text{and} \quad \begin{pmatrix} U \\ U \end{pmatrix}\ ,$$

whose rows of index $(r,1)$ et $(r,2)$ (for every r in R) are equal to the row of index r in $\mathcal{S}$. Hence, it holds:

$$|\mathcal{A}| = |\mathcal{S}| = |\mathcal{S}'| = |\langle \begin{pmatrix} J & J & 0 \end{pmatrix}, F', \begin{pmatrix} V \\ 0 \\ 0 \end{pmatrix} \rangle| + |\langle \begin{pmatrix} J & J & 0 \end{pmatrix}, F', \begin{pmatrix} 0 \\ W \\ U \end{pmatrix} \rangle|$$

and

$$|\mathcal{T}| = |\langle J, G, V \rangle| = |\langle \begin{pmatrix} J & J & 0 \end{pmatrix}, F', \begin{pmatrix} V \\ 0 \\ 0 \end{pmatrix} \rangle| = \underline{\mathsf{supp}\,|\mathcal{A}|}$$

The behaviour of the automaton

$$\mathcal{P} = \langle \begin{pmatrix} J & J & 0 \end{pmatrix}, F', \begin{pmatrix} 0 \\ W \\ U \end{pmatrix} \rangle$$

is then equal to $s - \underline{\mathsf{supp}\, s}$. ■

Example 2 (continued) : The above construction is applied to the automaton

$$\mathcal{C}_1 = \langle \begin{pmatrix} 1 & 0 \end{pmatrix}, \begin{pmatrix} a+b & b \\ 0 & 2a+2b \end{pmatrix}, \begin{pmatrix} 0 \\ 1 \end{pmatrix} \rangle\ .$$

This case is made simple by the fact that (with notation of the proof) $V = U$ and thus it holds directly that $\mathcal{S}' = \mathcal{S}''$ is a $\mathbb{N}$-covering of the Schützenberger covering of $\mathcal{C}_1$. The corresponding automaton $\mathcal{P}_1$ is drawn at the Figure 6. □

Theorem 1 yields directly a series of well-known corollaries that give useful insights on the structure of $\mathbb{N}$-rational series and that are worth recalling.

An $\mathbb{N}$-series is said to be *bounded (by k)* if the set of its coefficients is bounded (by k). Let s and t be two $\mathbb{N}$-series. We write $s \leqslant t$ if $<s,f> \leqslant <t,f>$ for every f in A^*, *i.e.* if there exists an $\mathbb{N}$-series u such that $s + u = t$; in this case we write $u = t - s$. More generally, the operation $t \dot{-} s$ is defined by

$$<t - s, f> = \mathsf{sup}\{0, (<t,f> - <s,f>)\}$$

for every f in A^*.

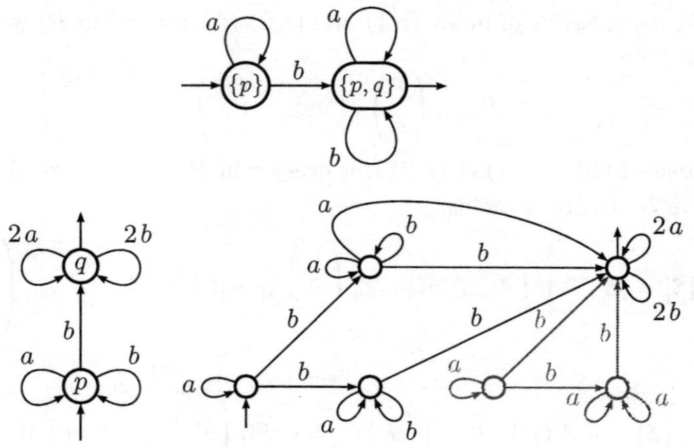

Figure 6: An automaton $\mathcal{P}_1$ whose behaviour is equal to $|\mathcal{C}_1| - \underline{\mathsf{supp}\,|\mathcal{C}_1|}$.

Corollary 7 *An $\mathbb{N}$-rational series bounded by k is the sum of at most k $\mathbb{N}$-rational characteristic series* ■

Corollary 8 *Let s and t be two $\mathbb{N}$-rational series such that s is bounded. Then $t \dot{-} s$ is an $\mathbb{N}$-rational series.* ■

Corollary 9 *Let s be an $\mathbb{N}$-rational series on A^*. For every integer k the languages*

$$\{f \in A^* \mid \langle s, f\rangle \geqslant k\}, \quad \{f \in A^* \mid \langle s, f\rangle = k\} \quad and \quad \{f \in A^* \mid \langle s, f\rangle \leqslant k\}$$

are rational. ■

Remark 2 The proof of Corollary 7 is indeed immediate: if $|\mathcal{A}|$ is bounded by k, we write $|\mathcal{A}| = |\mathcal{T}| + |\mathcal{P}|$ as above, $|\mathcal{P}|$ is bounded by $k - 1$ and we iterate the procedure. But it conceals a problem: at every step of that procedure, we have to perform a determinization (for the construction of the Schützenberger covering) which means an exponentiation. And this easy proof yields then a tower of k exponentiation. However, the work of Weber ([11]) on the decomposion of k valued transducers leads to think that a double exponentiation is sufficient in any case but this is still a conjecture.

Remark 3 The definition of $\mathbb{K}$-covering we have given as well as the construction of the automaton $\mathcal{P}$ in the proof of Theorem 1 may ring some bells to the reader who is familiar with symbolic dynamical system theory and who is reminiscent of the technic of *state splitting* and *state amalgamation* (*cf.* [3, §2.4] for instance).

If $\mathcal{B}$ is obtained from $\mathcal{A}$ by an In-splitting, then $\mathcal{B}$ is an $\mathbb{N}$-covering of $\mathcal{A}$ and, dually, $\mathcal{B}$ is an $\mathbb{N}$-co-covering of $\mathcal{A}$ if it is obtained by an Out-splitting. But the converse is not true. Roughly speaking, $\mathcal{B} = \langle J, F, U \rangle$ is an $\mathbb{N}$-covering of $\mathcal{A}$ if the rows of "equivalent" index of the matrix F' are equal, where F' is obtained from F by adding the columns of equivalent index. Whereas $\mathcal{B}$ is obtained from $\mathcal{A}$ by an In-splitting if the rows of "equivalent" index of F are equal (and thus they are equal in F').

Proposition 5 can then be seen as the equivalent of Williams' theorem in this setting.

Acknowledgements

It is a pleasure to thank again the organisers of the conference *Mathematical Foundations of Informatics 1999*, Prof. Do Long Van and Prof. Le Tuan Hoa, for their hospitality in Hanoï and for the beautiful job they did and to acknowledge Prof. Masami Ito's endless patience.

I wish to thank also Géraud Sénizergues who invited me to present the matter of this paper in a tutorial at the LANFOR symposium in March 99, and Pascal Weill who suggested me the name "skimming theorem".

References

[1] BERSTEL, J., AND REUTENAUER, C. *Les séries rationnelles et leurs langages.* Masson, 1984. Traduction: *Rational Series and their Languages.* Springer, 1986.

[2] EILENBERG, S. *Automata, Languages and Machines*, vol. A. Academic Press, 1974.

[3] LIND, D., AND MARCUS, B. *An introduction to symbolic dynamics and coding.* Cambridge University Press, 1995.

[4] SAKAROVITCH, J. A construction on automata that has remained hidden. *Theoret. Comput. Sci.* **204** (1998), 205–231.

[5] SAKAROVITCH, J. Eléments de théorie des automates. *in preparation.*

[6] Salomaa, A., and Soitttola, M. *Automata-Theoretic Aspects of Formal Power Series.* Springer, 1977.

[7] Schützenberger, M. P. Parties rationnelles d'un monoïde libre. *Proc. of the International Mathematics Conference* (1970), 281–282.

[8] Schützenberger, M. P. Sur les relations rationnelles entre monoïdes libres. *Theoret. Comput. Sci.* **3** (1976), 243–259.

[9] Schützenberger, M. P. Une propriété de Hankel des relations rationnelles entre monoïdes libres. *Advances in Math.* **24** (1977), 274–280.

[10] Stallings, J. Topology of finite graphs. *Inventiones Math.* **71** (1983), 551–565.

[11] Weber, A. Decomposing a k-valued transducer into k unambiguous ones. *Proc. of LATIN'92* (I. Simon, Ed.), Lecture Notes in Computer Sci. 583 (1992), 503–515.

A NEW CLASSIFICATION OF FINITE SIMPLE GROUPS

WUJIE SHI
Dept. of Math., Soochow University, Suzhou 215006 P.R. China and
Inst. of Math., Southwest China Normal University, Chongqing 400715 P.R. China
Email: wjshi@fm365.com

SEYMOUR LIPSCHUTZ
Dept. of Math., Temple University, Philadelphia PA 19122, USA
Email: Seymour@math.temple.edu

Let G be a finite simple group and $\pi_e(G)$ the set of element orders of G. For a subset Γ of all positive integers, let $h(\Gamma)$ denote the number of distinct groups G such that $\pi_e(G) = \Gamma$. For given group G, we have $h(\pi_e(G)) \geq 1$. In this paper we prove that $h(\pi_e(G)) \in \{1, 2, \infty\}$ for $|G| < 10^8$. Moreover we give a classification for such simple groups G.

1. Introduction and Notation

Element orders is one of the most fundamental concept like the **order of groups** in group theory. It plays an important role in the research of group theory, which can be seen from the famous Burnside problem. Let G be a group. Denote by $\pi_e(G)$ the set of all orders of elements in G. Obviously, $\pi_e(G)$ is a subset of the set Z^+ of positive integers, and it is a more difficult problem which subset of Z^+ can become the sets of element orders of groups. Let Γ be a subset of Z^+ and $h(\Gamma)$ be the number of isomorphism classes of groups G such that $\pi_e(G) = \Gamma$. For a given group G, we have $h(\pi_e(G)) \geq 1$. A group G is called **distinguishable** if $h(\pi_e(G)) < \infty$, a group G is called **non-distinguishable** if $h(\pi_e(G)) = \infty$, and a distinguishable group G is called ***k*-distinguishable** if $h(\pi_e(G)) = k$, moreover a 1-distinguishable group G is called **characterizible** usually. In [10] we prove that a finite distinguishable group is either characterizible or 2-distinguishable if its element orders do not exceed twenty. In this paper we continue this research and prove:

Theorem. Let G be a finite simple groups and $|G| < 10^8$. Then $h(\pi_e(G)) \in \{1, 2, \infty\}$.

Since we did not find such finite k-distinguishable groups for $k \geq 3$, we have the following conjecture:

Conjecture. Let G be a finite group. Then G is either characterizible, or 2-distinguishable, or non-distinguishable.

All groups considered in this paper are finite, a simple group is always nonabelian, and all notation is standard (see [2]). In particular, $\pi_e(G)$ denotes the set of all orders of elements in G and $\pi(G)$ denotes the set of all prime divisors of $|G|$. Moreover the prime graph ([25]) of a group G is the graph whose vertex set is the set $\pi(G)$, and two vertices $p, r \in \pi(G)$ are adjacent if and only if G contains an element of order pr. Also, a finite group G is called a 2-Frobenius groups if it has a normal series $G > K > H > 1$, where K is a Frobenius group with kernel H and G/H is a Frobenius group with kernel K/H.

2. Preliminary Lemmas

We will need the following lemmas.

Lemma 1. Let G be a group, and let N be a minimal normal subgroup of G. Suppose N is an elementary abelian p-group. Then G is non-distinguishable. In particular, a solvable group (finite or infinite) always is non-distinguishable.

Proof. See [22, Proposition] and [5, Theorem].

Lemma 2. Let p be a prime and b a positive integer. Then one of the following holds:

(1) There is a prime q such that $q \mid (p^b-1)$ but $q \nmid (p^c-1)$ for $c< b$.

(2) $b = 1$ or $p = 2$, $b = 6$.

(3) p is a Mersenne prime and $b = 2$.

Proof. See [26].

For the simple groups we have the following lemmas:

Lemma 3. Let G be a simple group and $|G| < 10^8$. Then G is one of the following groups:

(1) A_n, $n \le 11$.

(2) $L_2(q)$, $q \le 577$, $q \ne 2^9$; $L_3(q)$, $q \le 9$; $L_4(3)$, $L_5(2)$.

(3) $U_3(q)$, $q \le 11$; $U_4(2)$, $U_4(3)$, $U_5(2)$.

(4) $Sz(8)$, $Sz(32)$.

(5) $S_4(4)$, $S_4(5)$, $S_6(2)$.

(6) $G_2(3)$.

(7) ${}^2F_4(2)'$.

(8) M_{11}, M_{12}, M_{22}, M_{23}, J_1, J_2, J_3, HS.

Proof. By a calculation from the classification theorem of the finite simple groups we may get the result easily.

A finite simple group G will be called a simple K_n-group if $|\pi(G)| = n$. In [6] the simple K_3-groups are determined.

Lemma 4. Let G be a simple K_3-group. Then G is one of the following groups: A_5, A_6, $L_2(q)$, $q = 7, 8, 17$, $L_3(3)$, $U_3(3)$, $U_4(2)$.

For the above eight simple K_3-groups we have proved the following result.

Lemma 5. Let G be a simple K_3-group. Then we have

(1) $h(\pi_e(A_5)) = h(\pi_e(L_2(q))) = 1$, where $q = 7, 8, 17$.

(2) $h(\pi_e(A_6)) = h(\pi_e(L_3(3))) = h(\pi_e(U_3(3))) = h(\pi_e(U_4(2))) = \infty$.

Proof. See [18].

Lemma 6. Let G be a simple K_4-group. Then G is one of the following groups:

(1) A_n, $7 \le n \le 10$.

(2) M_{11}, M_{12}, J_2.

(3) $L_2(q)$, $q = 16, 25, 49, 81$, $L_3(q)$, $q = 4, 5, 7, 8, 17$, $L_4(3)$, $S_4(q)$, $q = 4, 5, 7, 9$, $S_6(2)$, $O^+_8(2)$, $G_2(3)$, $U_3(q)$, $q = 4, 5, 7, 8, 9$, $U_4(3)$, $U_5(2)$, $Sz(8)$, $Sz(32)$, ${}^3D_4(2)$, ${}^2F_4(2)'$.

(4) (a) $L_2(r)$, where r is a prime satisfying the equation

$$r^2 - 1 = 2^a 3^b u^c \tag{1}$$

where $a \ge 1$, $b \ge 1$, $c \ge 1$, and u is a prime and $u > 3$.

(b) $L_2(2^m)$, where m satisfies the equations

$$2^m - 1 = u,\ 2^m + 1 = 3t^b \tag{2}$$

where u, t are primes and $t > 3$ and $b \ge 1$.

(c) $L_2(3^m)$, where m satisfies the equations

$$3^m - 1 = 2u^c,\ 3^m + 1 = 4t \tag{3}$$

or

$$3^m - 1 = 2u,\ 3^m + 1 = 4t^b \tag{4}$$

where u, t are primes and $b \ge 1$, $c \ge 1$.

Proof. See [19].

Remark. It is difficult to solve these diophantine equations (1), (2) and (3) or (4), even determine the number (finite or infinite) of these solutions. That is, we do not know the number of the simple K_4–groups is finite, or infinite up to now.

Corollary 1. Let G be a simple K_4-group and $|G| = 2^a 3^b 5^c 13^d$. Then G is one of the following groups: $L_2(25)$, $L_4(3)$, $S_4(5)$, $U_3(4)$, ${}^2F_4(2)'$.

Corollary 2. Let G be a simple K_4-group and $|G| = 2^a 3^b 7^c 19^d$. Then G is one of the following groups: $L_3(7)$, $U_3(8)$.

Corollary 3. Let G be a simple K_4-group and $|G| = 2^a 3^b 5^c 73^d$. Then G is $U_3(9)$.

Corollary 4. Let G be a simple K_4-group and $|G| = 2^a 3^b 5^c 31^d$. Then G is one of the following groups: $L_2(31)$, $L_3(5)$.

Corollary 5. There is no existing such simple K_4-group G with the order of G is one of the following forms: $2^a3^b7^c31^d$, $2^a5^b7^c31^d$, $2^a3^b5^c37^d$, $2^a3^b11^c37^d$, $2^a5^b11^c37^d$.

Lemma 7. If G is a simple group whose order is divisible by 31 and $\pi_e(G) \subseteq \pi_e(L_5(2))$, then G is $L_5(2)$.

Proof. The order of G is equal to $2^a3^b5^c7^d31^e$, where $e \neq 0$. From Lemma 4, Corollary 4 and Corollary 5 we see that G can only be a simple K_5-group (since $16 \in \pi_e(L_2(31))$, $16 \notin \pi_e(L_5(2))$ and $10 \in \pi_e(L_3(5))$, $10 \notin \pi_e(L_5(2))$). Since $(n!)/2$ is divisible by 31 if and only if $n \geq 31$, G is not an alternating group A_n for any n. Comparing the order of G with the orders of sporadic simple groups we have G is not a sporadic group. Therefore $G = G(q)$ is a group of Lie type over a finite field of order $q = p^n$, where $p = 2, 3, 5, 7$ or 31, since p divides $|G|$. From Lemma 2 we get that $p^b - 1$ can divide $|G|$ if and only if:

for $p = 2$, $b \leq 6$,
for $p = 3$ or 7, $b = 1,2$ and 4,
for $p = 5$, $b = 1, 2, 3$ and 6,
for $p = 31$, $b \leq 2$.

Also $p^b + 1$ can divide $|G|$ if and only if:

for $p = 2$ or 3, $b \leq 3$,
for $p = 5$, $b = 1$ and 3,
for $p = 7$, $b \leq 2$,
for $p = 31$, $b = 1$.

Moreover G can only be $L_2(125)$, $L_5(2)$, $L_6(2)$ or $G_2(5)$ since G is a simple K_5-group. Because of $63 \in \pi_e(L_2(125))$, $63 \in \pi_e(L_6(2))$, $63 \notin \pi_e(L_5(2))$ and $30 \in \pi_e(G_2(5))$, $30 \notin \pi_e(L_5(2))$, G can only be $L_5(2)$.

Lemma 8. If G is a simple group whose order is divisible by 37 and $\pi_e(G) \subseteq \pi_e(U_3(11))$, then G is $U_3(11)$.

Proof. Notice that the order of G is equal to $2^a3^b5^c11^d37^e$, $e \neq 0$, G can only be a simple K_5-group by Lemma 4 and Corollary 6. The remainder proof is similar to Lemma 7.

For some simple groups and nonsolvable groups we have the following results:

Lemma 9. The following groups are all characterizible:

(1) Alternating groups A_p, p is a prime and $p > 3$, or A_n for $n \in \{8, 9, 12, 14\}$; symmetric groups S_n for $n \in \{7, 9, 11, 12, 13, 14\}$.

(2) Simple groups $L_2(q)$ with $q \neq 9$ ($L_2(9) \cong A_6$), $L_2(9).2_3$ ($\cong M_{10}$); $L_3(7)$, $L_3(2^m)$ with $m \geq 1$, $L_3(4).2_1$; $L_4(3)$.

(3) $U_4(3)$, $U_6(2)$, $U_3(2^m)$ with $m \geq 2$.

(4) Suzuki-Ree simple groups.

(5) $G_2(3)$.

(6) ${}^2F_4(2)'$.

(7) $O^-_8(2)$, $O^-_{10}(2)$.

(8) All sporadic simple groups except J_2.

Proof. See [20], [12], [10], [14], [1], [22], [15], [4] and [9].

A group G is called an almost simple group if $M \le G \le Aut(M)$, where M is a simple group.

Lemma 10. The following almost simple groups are non-distinguishable: A_n for $n \in \{6, 10\}$, S_n for $n \in \{5, 6, 8\}$, $L_3(3)$, $U_3(q)$ with $q = 3, 5, 7$, $U_4(2)$, $U_5(2)$, J_2, $L_3(4) : 2_2$, $S_4(7)$, $S_4(2^m)$ with $m > 1$, and $PGL_p(r^s)$, where p, r are odd primes, $r - 1$ is divisible by p but not by p^2 and s is a natural number non-divisible by p.

Proof. See [14], [13] and [15].

Lemma 11. The following groups are 2-distinguishable:

(1) $L_3(5)$, $L_3(5).2$ ($\cong Aut(L_3(5))$).

(2) $L_3(9)$, $L_3(9).2_1$.

(3) $S_6(2)$, $O^+_8(2)$.

(4) $O_7(3)$, $O^+_8(3)$.

Proof. See [11], [12], [3] and [24].

Lemma 12. Let G be a finite group and $\pi_e(G) \subseteq \{1, 2, \ldots, 20\}$. Then $h(\pi_e(G)) \in \{1, 2, \infty\}$.

Proof. See [10].

Lemma 13. Let H be a finite group and N a nontrivial normal p-subgroup of H, for some prime p, and set $K = H/N$. Suppose K contains an element x of order m coprime to p such that $\langle \varphi|_{\langle x\rangle}, 1_{\langle x\rangle}\rangle > 0$ for every Brauer character φ of (an absolutely irreducible representation of) K in characteristic p, where $1_{\langle x\rangle}$ is the trivial character of $\langle x\rangle$ and $\varphi|_{\langle x\rangle}$ is the restriction of φ to $\langle x\rangle$. Then H contains an element of order pm.

Proof. See [17, Corollary].

3. Main Results

From Lemma 1, we may reduce the proof of our Theorem to the direct products of simple groups and the automorphism groups of the direct products. First we prove the following theorem.

Theorem 1. Let G be one of the following groups: $S_4(5)$, $L_5(2)$, $U_3(9)$, $U_3(11)$. Then $h(\pi_e(G)) \in \{1, \infty\}$.

Proof. Let H be a group and suppose $\pi_e(H) = \pi_e(G)$. If the minimal normal subgroup N of H is a p-group, then $h(\pi_e(H)) = h(\pi_e(G)) = \infty$ by Lemma 1. Thus we may assume that H has a normal series

$$H \geq H_1 > 1, \tag{5}$$

where H_1 is a direct product of isomorphic simple groups. Since H is a group whose prime graph has more than one component, H has one of the following structures: (a) Frobenius or 2-Frobenius; (b) simple; (c) an extension of a π_1-group by a simple group; (d) simple by π_1-solvable; or (e) π_1 by simple by π_1; where π_1 is the component containing 2 ([25, Theorem A]). Suppose H is a Frobenius group with complement C. From [16, Theorem 18.6] C has a normal subgroup C_0 of index ≤ 2 such that $C_0 \cong SL(2, 5) \times Z$ and where every Sylow subgroup of Z is cyclic and $\pi(Z) \cap \{2,3,5\} = \varnothing$. Since $\pi_e(SL(2,5)) = \{1, 2, \ldots, 6, 10\}$, we get a contradiction by comparing their element orders. The case when H is 2-Frobenius is similar. Therefore H_1 is a simple group and H/H_1 is a π_1-group in (5). Now we prove that $H \cong G$ if $h(\pi_e(H)) \neq \infty$. The proof is divided into cases.

Case 1. $G = S_4(5)$

In this case we have $\pi_e(H) = \pi_e(G) = \pi_e(S_4(5)) = \{1, 2, \ldots, 6, 10, 12, 13, 15, 20, 30\}$ and $\pi_e(H_1) \subseteq \pi_e(H)$. Hence H_1 is a simple group whose order is divisible by 13. From $p^a q^b$ theorem H_1 is a simple K_3-group or a simple K_4-group. If H_1 is a simple K_3-group, then H_1 can only be $L_3(3)$ ($13 \mid |H_1|$). But $L_3(3)$ contains the elements of order 8, it is impossible. If H_1 is a simple K_4-group, then H_1 can only be $L_2(25)$, $U_3(4)$ or $S_4(5)$ from Corollary 2 and comparing their element orders.

Now $H/C_H(H_1)$ is isomorphic to a subgroup of $Aut(H_1)$, and $C_H(H_1) = 1$ (H_1 has at least two components). We have $Aut(H_1) \geq H \geq H_1$. Since we know $Out(H_1)$ and check the orders of elements of $Aut(H_1)$ one by one (see [2]), we get $H = H_1 \cong G = S_4(5)$.

Case 2. $G = L_5(2)$

In this case we have $\pi_e(H) = \pi_e(G) = \{1, 2, \ldots, 8, 12, 14, 15, 21, 31\}$ and $\pi_e(H_1) \subseteq \pi_e(H)$. Since H_1 is a simple group whose order is divisible by 31, H_1 can only be $L_5(2)$ by Lemma 7. Also since $C_H(H_1) = 1$ we have $Aut(H_1) \geq H \geq H_1$. Checking the orders of elements of $Aut(L_5(2))$ we get $H = H_1 \cong G = L_5(2)$.

Case 3. $G = U_3(9)$

In this case we have $\pi_e(H) = \pi_e(G) = \{1, 2, \ldots, 6, 8, 10, 15, 16, 20, 30, 40, 73, 80\}$ and $\pi_e(H_1) \subseteq \pi_e(H)$. Since H_1 is a simple group whose order is divisible by 73, H_1 can only be $U_3(9)$ from Lemma 4 and Corollary 3. Also since $C_H(H_1) = 1$ we have $Aut(H_1) \geq H \geq H_1$. Checking the orders of elements of $Aut(U_3(9))$ we get $H = H_1 \cong G = U_3(9)$.

Case 4. $G = U_3(11)$

In this case we have $\pi_e(H) = \pi_e(G) = \{1, 2, \ldots, 6, 8, 10, 11, 12, 20, 22, 37, 40, 44\}$ and $\pi_e(H_1) \subseteq \pi_e(H)$. Since H_1 is a simple group whose order is divisible by 37, H_1 can only be $U_3(11)$ from Lemma 8. Also since $C_H(H_1) = 1$ we have $Aut(H_1) \geq H \geq H_1$. Checking the orders of elements of $Aut(U_3(11))$ we get $H = H_1 = G = U_3(11)$.

Thus the theorem is proved.

Theorem 2. The simple groups $L_5(2)$ and $U_3(11)$ are characterizible, and $S_4(5)$ is non-distinguishable.

Proof. From Theorem 1 we need only prove the minimal normal p-subgroup N of H is equal to 1 for the cases of $L_5(2)$ and $U_3(11)$. Let H be a group and suppose $\pi_e(H) = \pi_e(G)$, $G = L_5(2)$ or $U_3(11)$. First we prove that H is nonsolvable. We assume $G = L_5(2)$ ($U_3(11)$). Then $\pi_e(H) = \pi_e(G) = \{1,2, \ldots ,8,12,14,15,21,31\}$ ($\{1,2, \ldots ,6,8,10,11,12,20,22,37,40,44\}$). Suppose H were solvable. Then there would exists a $\{5, 7, 31\}$ ($\{5, 11, 37\}$)-Hall subgroup L of H. Now $5\cdot 7$, $5\cdot 31$ and $7\cdot 31$ ($5\cdot 11$, $5\cdot 37$ and $11\cdot 37$) do not belong to $\pi_e(H)$, and so all elements of L would be of prime power order. This contradicts a result of Higman [7], and hence H is nonsolvable. Similar to the proof of Theorem 1 we may assume that H has a normal series

$$H \geq H_1 > N \geq 1$$

such that N and H/H_1 are π_1-groups, and $H^* = H_1/N$ is simple. Now we need only prove $N = 1$ from Theorem 1.

Case 1. $G = L_5(2)$

Since $\pi_1 = \{2, 3, 5, 7\}$ we may assume that $N \neq 1$ is a 2-group, 3-group, 5-group or 7-group. Suppose N is a 2-group (5-group or 7-group). Let Q be a cyclic subgroup of order 31 in H^*, and $N(Q) = N_{H^*}(Q)$. By [2], $N(Q) = Q{:}Z_5$ and suppose $[N](Q{:}Z_5)$ is its preimage in H_1. It follows by [21, Lemma 6] that H_1 contains an element of order 10 (25 or 35), which is impossible. Suppose N is a 3-group. By Lemma 13 we utilize the Brauer character tables. In [8, p.173], the modular characters of $L_5(2)$ (mod 3) are given. Suppose $x \in H^*$ has order 8, and set $X = \langle x \rangle$. Calculating, we obtain

$$\langle \varphi|_X, 1|_X \rangle = \frac{1}{|X|}\sum_{x\in X} \varphi(x) > 0.$$

Thus by Lemma 13 we get $24 \in \pi_e(H)$, which is a contradiction. Hence $N = 1$ and $H = G$. The simple group $L_5(2)$ is characterizible.

Case 2. $G = U_3(11)$

Since $\pi_1 = \{2, 3, 5, 11\}$ we may assume that $N \neq 1$ is a 2-group, 3-group, 5-group or 11-group. Suppose N is a 3-group (5-group or 11-group). Let Q be a cyclic subgroup of order 37 in H^*, and $N(Q) = N_{H^*}(Q)$. By [2], $N(Q) = Q{:}Z_3$ and suppose $[N](Q{:}Z_3)$ is its preimage in H_1. It follows by [21, Lemma 6] that H_1 contains an element of order 9 (15 or 33), which is impossible. Suppose N is a 2-group. We utilize the Brauer character tables [8, p.220] and set $X = \langle x \rangle$, $|x| = 37$. Calculating, we obtain $74 \in \pi_e(H)$ by Lemma 13, which is a contradiction. Hence $N = 1$, $H = G$ and $U_3(11)$ is characterizible.

Case 3. $G = S_4(5)$

If $G = S_4(5)$, then $\pi_e(G) = \pi_e(S_4(5)) = \{1, 2, \ldots, 6, 10, 12, 13, 15, 20, 30\}$. The group $5^4{:}\ (L_2(25).2_2)$ which is contained in the maximal 5-local subgroups of Monster group M has the same element orders set (see [2]). Since the minimal normal subgroup is elementary abelian, we get that $S_4(5)$ is non-distinguishable. This proves Theorem 2.

Lastly, we prove the following theorem.

Theorem. Let G be a finite simple group and $|G| < 10^8$. Then $h(\pi_e(G)) = \{1, 2, \infty\}$.

Proof. Let G be a simple group and $|G| < 10^8$. Then G is either characterizible, or 2-distinguishable, or non-distinguishable by Lemma 3, Theorem 1, Theorem 2, Lemma 5, Lemma 9, Lemma 10 and Lemma 11. Moreover, the following simple groups G ($|G| < 10^8$) are characterizible:

(1) A_n, $n \leq 11$ and $n \neq 6, 10$.

(2) $L_2(q)$, $2^9 \neq q \leq 577$ and $q \neq 9$, $L_3(4)$, $L_3(7)$, $L_4(3)$, $L_5(2)$.

(3) $U_3(4)$, $U_3(8)$, $U_3(11)$, $U_4(3)$.

(4) $Sz(8)$, $Sz(32)$.

(5) $G_2(3)$.

(6) ${}^2F_4(2)'$.

(7) M_{11}, M_{12}, M_{22}, M_{23}, J_1, J_3, HS.

The following simple groups G are 2-distinguishable:

(1) $L_3(5)$, $L_3(9)$.

(2) $S_6(2)$.

The following simple groups G are non-distinguishable:

(1) A_6, A_{10}.

(2) $L_2(9)$, $L_3(3)$.

(3) $U_3(3)$, $U_3(5)$, $U_3(7)$, $U_4(2)$, $U_5(2)$.

(4) $S_4(4)$ (see [10]), $S_4(5)$.

(5) J_2.

Excluding the above cases, we have $h(\pi_e(U_3(9))) \in \{1, \infty\}$. Thus the theorem is proved.

Question. For these case that G is $U_3(9)$, does it follow that $h(\pi_e(G)) = 1$ or $h(\pi_e(G)) = \infty$?

Acknowledgments

The present paper was written while the first author visited Temple University in 1998. Thanks are due to Temple University for its hospitality and to the National Natural Science Foundation of China (Grant No. 10171074), Jiangsu Natural Science Foundation (Grant No. BK200133) and the NSF of USA (Grant INT 9600217) for financial support.

Reference

1. J An and Wujie Shi, *Comm. in Algebra* 28, 3351(2000).
2. J H Conway et al, *ATLAS of Finite Groups*, (Clarendon Press, Oxford, 1985).
3. N Chigira and Wujie Shi, *Northeast Math. J.* 12, 253(1996).
4. M R Darafsheh and A R Moghaddamfar, *Algebra Colloquium* 7, 467 (2000).
5. Huiwen Deng and Wujie Shi, *J. Southwest-China Normal Univ.* 25, 361 (2000).
6. M Herzog, *J. Algebra* 10, 383(1968).
7. G Higman, *J. London Math. Soc.* 32, 335(1957).
8. C Jansen et al, *An Atlas of Brauer Characters*, (Clarendon Press, Oxford, 1995).
9. A S Kondrat'ev and V D Mazurov, *Siberian Math. J.* 41, 360(2000).
10. S Lipschutz and Wujie Shi, *Progress in Natural Science* 10, 11(2000).
11. V D Mazurov, *Algebra and Logic* 33, 49(1994).
12. V D Mazurov, *Algebra and Logic* 36, 23(1997).
13. V D Mazurov, *Algebra and Logic* 37, 371(1998).
14. V D Mazurov and Wujie Shi, in *St Andrews 1997 in Bath, II, London Math. Soc. Lecture Notes Ser. 261*, eds. C M Campbell et al, (Cambridge University Press, 1999).
15. V D Mazurov, M C Xu and H P Cao, *Algebra and Logic* 39, 324(2000).
16. D S Passman, *Permutation Groups*, (W.A. Benjamin, New York, 1968).
17. C E Prager and Wujie Shi, *Comm. in Algebra* 22, 1507(1994).
18. Wujie Shi, *J. Southwest-China Normal Univ.* 13, 1(1988).
19. Wujie Shi, *Chinese Sci. Bull.* 36, 1281(1991).
20. Wujie Shi, in *Group Theory in China, Math. And Its Appl.* (*China Ser.*), eds. Z X Wan and S M Shi, (Sci. Press, New York/Beijing, 1996).

21. Wujie Shi, *Algebra Colloquium* 1, 159(1994).
22. Wujie Shi, *J. Southwest-China Normal Univ.* (*Suppl. 1*) 21, 6(1996).
23. Wujie Shi, *J. Southwest-China Normal Univ.* 25, 353(2000).
24. Wujie Shi and C Y Tang, *Progress in Natural Science* 7, 155(1997).
25. J S Williams, *J. Algebra* 69, 487(1981).
26. K Zsigmondy, *Monatsh. Math. Phys.* 3, 265(1892).

CONNECTEDNESS OF TETRAVALENT METACIRCULANT GRAPHS WITH NON-EMPTY FIRST SYMBOL

Ngo Dac Tan* and Tran Minh Tuoc†

Abstract

In this paper, we give necessary and sufficient conditions for tetravalent metacirculant graphs, the first symbol of which is not empty, to be connected.

1 Introduction

Metacirculant graphs were introduced by Alspach and Parsons in [1] as an interesting class of vertex-transitive graphs. This class contains both Cayley and non-Cayley graphs [1] and might contain some new nonhamiltonian connected vertex-transitive graphs. But these graphs need not be connected in general. So a natural problem is to find necessary and sufficient conditions for metacirculant graphs to be connected.

Necessary and sufficient conditions for cubic metacirculant graphs to be connected were found in [4]. These conditions were used successfully in [3, 5] to prove the existence of a Hamilton cycle in many connected cubic metacirculant graphs (see also [6]). Motivated by this, we consider here the connectedness of tetravalent metacirculant graphs with the hope that we might apply the obtained results to the hamiltonian problem for tetravalent metacirculant graphs as well. In Section 4 of this paper, we will prove necessary and sufficient conditions for tetravalent metacirculant graphs with non-empty first symbol to be connected. For the proofs of the results, we develop in Section

*Institute of Mathematics, 18 Hoang Quoc Viet Road, 10307 Hanoi, Vietnam. E-mail: ndtan@math.ac.vn

†Faculty of Mathematics, Pedagogic University Hanoi 2, Xuanhoa, Melinh, Vinhphuc, Vietnam. E-mail: minhtuoc@hn.vnn.vn

3 general techniques which are probably useful in other cases. In particular, they were applied in [8] to prove necessary and sufficient conditions for tetravalent metacirculant graphs with empty first symbol to be connected. Thus, the results obtained here and in [8] together gave us a complete answer to the question which tetravalent metacirculant graphs are connected.

We also note that with the help of the results obtained here the authors have succeeded in proving the existence of a Hamilton cycle in many connected tetravalent metacirculant graphs with nonempty first symbol [7]. We do hope that using the mentioned results we will get more significant results on the existence of a Hamilton cycle in tetravalent metacirculants.

2 Notations and definitions

All graphs considered in this paper are finite undirected graphs without loops and multiple edges. Unless otherwise indicated, our graph-theoretic terminology will follow [2], and our group-theoretic terminology will follow [9]. For a graph G we will denote by $V(G), E(G)$ and $\mathrm{Aut}(G)$ the vertex-set, the edge-set and the automorphism group of G, respectively.

A graph G is called *vertex-transitive* if for any $u, v \in V(G)$ there exists $\varphi \in \mathrm{Aut}(G)$ such that $\varphi(u) = v$. It is clear that a vertex-transitive graph is a regular graph.

For a positive integer n, we will denote the ring of integers modulo n by $\mathbb{Z}_n$ and the multiplicative group of units in $\mathbb{Z}_n$ by $\mathbb{Z}_n^*$.

Let n be a positive integer and S be a subset of $\mathbb{Z}_n$ such that $0 \notin S = -S$. Then we define the *circulant graph* $G = C(n, S)$ to be the graph with vertex-set $V(G) = \{v_y \mid y \in \mathbb{Z}_n\}$ and edge-set $E(G) = \{v_y v_h \mid y, h \in \mathbb{Z}_n;\ (h - y) \in S\}$, where subscripts are always reduced modulo n. The subset S is called the *symbol* of $C(n, S)$.

Let m, n be two positive integers, $\alpha \in \mathbb{Z}_n^*, \mu = \lfloor m/2 \rfloor$ and $S_0, S_1, \ldots, S_\mu$ be subsets of $\mathbb{Z}_n$, satisfying the following conditions:

1) $0 \notin S_0 = -S_0$;

2) $\alpha^m S_r = S_r$ for $0 \leq r \leq \mu$;

3) If m is even, then $\alpha^\mu S_\mu = -S_\mu$.

Then we define the *metacirculant graph* $G = MC(m, n, \alpha, S_0, \ldots, S_\mu)$ to be the graph with vertex-set

$$V(G) = \{v_j^i \mid i \in \mathbb{Z}_m;\ j \in \mathbb{Z}_n\}$$

and edge-set

$$E(G) = \{v_j^i v_h^{i+r} \mid 0 \le r \le \mu;\ i \in \mathbb{Z}_m;\ j, h \in \mathbb{Z}_n \text{ and } (h - j) \in \alpha^i S_r\},$$

where superscripts and subscripts are always reduced modulo m and modulo n, respectively. The subset S_i is called $(i+1)$*-th symbol* of G.

It is easy to see that the permutations ρ and τ on $V(G)$ with $\rho(v_j^i) = v_{j+1}^i$ and $\tau(v_j^i) = v_{\alpha j}^{i+1}$ are automorphisms of G and the subgroup $\langle \rho, \tau \rangle$ generated by ρ and τ is a transitive subgroup of $\mathrm{Aut}(G)$. Thus, metacirculants are vertex-transitive graphs and therefore they are regular.

Denote the degree of vertex v of a graph G by $\deg(v)$. It is not difficult to see that for any vertex $v \in V(G)$ of a metacirculant graph $G = MC(m, n, \alpha, S_0, \ldots, S_\mu)$

$$\deg(v) = \begin{cases} |S_0| + 2|S_1| + \cdots + 2|S_\mu| & \text{if } m \text{ is odd,} \\ |S_0| + 2|S_1| + \cdots + 2|S_{\mu-1}| + |S_\mu| & \text{if } m \text{ is even.} \end{cases} \tag{E.1}$$

A graph G is called *cubic* if for any $v \in V(G)$, $\deg(v) = 3$ and it is called *tetravalent* if for any $v \in V(G)$, $\deg(v) = 4$. If $W \subseteq V(G)$ then we denote the subgraph induced by G on W by $G[W]$.

Let $W = v_{j_1}^{i_1} v_{j_2}^{i_2} \ldots v_{j_t}^{i_t}$ be a walk in $G = MC(m, n, \alpha, S_0, \ldots, S_\mu)$. Then the value $(j_t - j_1)$ modulo n is called the *change* (in subscripts) of W and is denoted by $ch(W)$. The walk $W^{-1} = v_{j_t}^{i_t} \ldots v_{j_2}^{i_2} v_{j_1}^{i_1}$ is called the inverse walk of W. Let $U = v_{j_t}^{i_t} v_{j_{t+1}}^{i_{t+1}} \ldots v_{j_s}^{i_s}$ be another walk in G, which starts at the vertex where W terminates. Then the walk $P = v_{j_1}^{i_1} v_{j_2}^{i_2} \ldots v_{j_t}^{i_t} v_{j_{t+1}}^{i_{t+1}} \ldots v_{j_s}^{i_s}$ is called the *concatenation* of W and U and is denoted by $W * U$. It is easy to see that concatenation operation of walks is associative, i. e. , $(W_1 * W_2) * W_3 = W_1 * (W_2 * W_3)$. Further, we have $ch(W^{-1}) \equiv -ch(W) \pmod{n}$, $ch(W * U) \equiv ch(W) + ch(U) \pmod{n}$ and if a walk W has the form $W = W_1 * Q * Q^{-1} * W_2$ then $ch(W) = ch(W_1 * W_2) \pmod{n}$.

Let $G = MC(m, n, \alpha, S_0, \ldots, S_\mu)$ be a metacirculant graph. Then an edge e of G is called an S_i^+*-edge* if it is $v_y^x v_h^{x+i}$ with $(h - y) \in \alpha^x S_i$ and an S_i^-*-edge* if it is $v_y^x v_h^{x-i}$ with $(y - h) \in \alpha^{x-i} S_i$. Both an S_i^+-edge and an S_i^--edge are called S_i*-edges*. If all edges of a walk W are S_i^+-edges (resp. S_i^--edges, S_i-edges) then W is called an S_i^+*-walk* (resp. S_i^-*-walk*, S_i*-walk*). A maximal S_i^+-subwalk (resp. S_i^--subwalk, S_i-subwalk) of W is called an S_i^+*-interval* (resp. S_i^-*-interval*, S_i*-interval*) of W. A subwalk W' of a walk W is called an *interval* of W if it is an S_i-interval for some $i \in \{0, 1, \ldots, \mu\}$. So each walk W in G can be represented in the form $W = W_1 * W_2 * \cdots * W_k$, where $W_1, W_2, \ldots, W_k$ are intervals of W.

3 General reduction results

In this section, we will prove some general results concerning connectedness of metacirculant graphs.

Let $G = MC(m, n, \alpha, S_0, \ldots, S_\mu)$ be a metacirculant graph. Denote $V^i = \{v_j^i \mid j \in \mathbb{Z}_n\}$. We define graphs $\overline{G}$ and G^i as follows. The graph $\overline{G}$ has the vertex-set $V(\overline{G}) = \overline{V} = \{V^0, V^1, \ldots, V^{m-1}\}$ and the edge-set $E(\overline{G}) = \overline{E} = \{V^iV^j \mid i \neq j$ and there exists $v_p^iv_q^j \in E(G)$ for some $p, q \in \mathbb{Z}_n\}$. The graph G^i has the vertex-set $V(G^i) = V^i$ and the edge-set $E(G^i) = E^i = \{v_k^iv_\ell^i \mid k \neq \ell$ and there exists a walk in G joining v_k^i to $v_\ell^i\}$, $i \in \{0, 1, \ldots, m-1\}$.

Lemma 3.1. *Let $G = MC(m, n, \alpha, S_0, \ldots, S_\mu)$ be a metacirculant graph. Then the following assertions hold:*

1. *$\overline{G}$ is isomorphic to $C(m, \overline{S})$, where $\overline{S} = \{h \in \mathbb{Z}_m \mid V^0V^h \in \overline{E}\}$.*
2. *G^i is isomorphic to $C(n, S^i)$, where $S^i = \{j \in \mathbb{Z}_n \mid v_0^iv_j^i \in E^i\}$.*
3. *All graphs G^i, $i \in \mathbb{Z}_m$, are isomorphic to each other.*

Proof. (1) Let $\overline{S} = \{h \in \mathbb{Z}_m \mid V^0V^h \in \overline{E}\}$. Then $0 \notin \overline{S}$. If $h \in \overline{S}$ then there is an edge $v_p^0v_q^h \in E(G)$ for some $p, q \in \mathbb{Z}_n$. It is easy to see that the permutation $\tau : V(G) \to V(G)$, $v_j^i \mapsto v_{\alpha j}^{i+1}$ is an automorphism of G. So $\tau^{-h} \in \mathrm{Aut}(G)$. Therefore $\tau^{-h}(v_p^0v_q^h) = \tau^{-h}(v_p^0)\tau^{-h}(v_q^h) = v_{p'}^{-h}v_{q'}^0$, where $p' \equiv \alpha^{-h}p \pmod n$, $q' \equiv \alpha^{-h}q \pmod n$, is an edge of G. This means $-h \in \overline{S}$ and $\overline{S} = -\overline{S}$. Let $\varphi : V(\overline{G}) \to V(C(m, \overline{S}))$, $V^i \mapsto v_i$ $(i \in \mathbb{Z}_m)$. Then it is not difficult to verify that φ is an isomorphism between $\overline{G}$ and $C(m, \overline{S})$.

(2) Let $G^i = (V^i, E^i)$ and $S^i = \{s \in \mathbb{Z}_n \mid v_0^iv_s^i \in E^i\}$. Then $0 \notin S^i$. Assume that $s \in S^i$. Then there is a walk $W = v_0^iv_{j_1}^{i_1}\ldots v_{j_f}^{i_f}v_s^i$ in G, which joins v_0^i to v_s^i. It is easy to verify that the permutation $\rho : V(G) \to V(G)$, $v_j^i \mapsto v_{j+1}^i$ is an automorphism of G. So $\rho^{-s} \in \mathrm{Aut}(G)$. Therefore $\rho^{-s}(W^{-1}) = \rho^{-s}(v_s^iv_{j_f}^{i_f}\ldots v_{j_1}^{i_1}v_0^i) = \rho^{-s}(v_s^i)\rho^{-s}(v_{j_f}^{i_f})\ldots\rho^{-s}(v_{j_1}^{i_1})\rho^{-s}(v_0^i) = v_0^iv_{j_f-s}^{i_f}\ldots v_{j_1-s}^{i_1}v_{-s}^i$ is also a walk in G. This means $-s \in S^i$ and $S^i = -S^i$. Let $\psi : V(G^i) \to V(C(n, S^i))$, $v_j^i \mapsto v_j$. Then it is easy to see that ψ is an isomorphism between G^i and $C(n, S^i)$.

(3) Consider the graphs G^k and G^ℓ. Since $\tau^{\ell-k} \in \mathrm{Aut}(G)$, where τ is as above, it is not difficult to verify that the restriction of $\tau^{\ell-k}$ on V^k is an isomorphism of G^k and G^ℓ. □

By this lemma, we can identify $\overline{G}$ with $C(m, \overline{S})$ and G^i with $C(m, S^i)$.

Lemma 3.2. *Let $G = MC(m, n, \alpha, S_0, \dots, S_\mu)$ be a metacirculant graph. Then G is connected if and only if both $\overline{G}$ and G^0 are connected.*

Proof. If G is connected, then it is clear that both $\overline{G}$ and G^0 are connected. Conversely, if both $\overline{G}$ and G^0 are connected. Then by Lemma 3.1, the graphs $G^1, \dots, G^{m-1}$ are also connected. Let v_ℓ^k be any vertex of G. Since $\overline{G}$ is connected, we can find a walk $v_0^0 v_{j_1}^{i_1} \dots v_j^k$ in G joining v_0^0 to a vertex v_j^k of V^k. Since G^k is connected, there is a walk in G joining v_j^k to v_ℓ^k. By the concatenating these walks, we can connect v_0^0 to v_ℓ^k in G. From this and the vertex-transitivity of G, it follows that G is connected. □

Lemma 3.3 ([4]). *Let $G = C(n, S)$ be a circulant graph with the symbol $S = \{\pm s_1, \dots, \pm s_k\}$. Then G is connected if and only if $gcd(s_1, \dots, s_k, n) = 1$.* □

Let $G = C(n, S)$ be a circulant graph and R be a subset of S satisfying the following conditions:

(i) $R = -R$;

(ii) For every $s \in S$, we can write $s = \sum_{i=1}^h t_i r_i$, where $t_i \in \mathbb{Z}$, $r_i \in R$.

Then we say that S is generated by R and denote this fact by $S = \langle R \rangle$.

Lemma 3.4. *Let $G = C(n, S)$ be a circulant graph with $S = \langle R \rangle$. Then G is connected if and only if $C(n, R)$ is connected.*

Proof. Let $G' = C(n, R)$. Since $R \subseteq S$, it is clear that connectedness of G' implies the connectedness of G. Conversely, let G be connected and v_k be any vertex of G. Then there exist $s_1, \dots, s_q \in S$ such that $v_0 v_{s_1}, v_{s_1} v_{s_1+s_2}, \dots, v_{s_1+s_2+\dots+s_{q-1}} v_k$ are in $E(G)$, where $k = s_1 + s_2 + \dots + s_q$. By Condition (ii), $s_i = \sum_{j=1}^{h_i} t_{i_j} r_{i_j}$, $t_{i_j} \in \mathbb{Z}$, $r_{i_j} \in R$, $i = 1, 2, \dots, q$. Therefore $k = \sum_{j=1}^h \alpha_j r_j$ for suitable values $h, \alpha_j \in \mathbb{Z}$ and $r_j \in R$, $j = 1, 2, \dots, h$. Now we can join v_0 to v_k by a walk in G' as follows: v_0 is joined to $v_{\alpha_1 r_1}$ by the walk $v_0 v_{r_1} v_{2r_1} \dots v_{\alpha_1 r_1}$; $v_{\alpha_1 r_1}$ is joined to $v_{\alpha_1 r_1 + \alpha_2 r_2}$ by the walk $v_{\alpha_1 r_1} v_{\alpha_1 r_1 + r_2} v_{\alpha_1 r_1 + 2r_2} \dots v_{\alpha_1 r_1 + \alpha_2 r_2}$, and so on. From this and the vertex-transitivity of G', it follows that G' is connected. □

Lemma 3.5. *A metacirculant graph $G = MC(m, n, \alpha, S_0, \dots, S_\mu)$ with $S_0 \neq \emptyset$ is tetravalent if and only if one of the following cases holds:*

1. *$|S_0| = 4$ and $S_1 = \dots = S_\mu = \emptyset$;*
2. *m and n are even, $|S_0| = 3$, $S_j = \emptyset$ for any $j \in \{1, 2, \dots, \mu - 1\}$ and $|S_\mu| = 1$;*

3. *m is even,* $|S_0| = 2$*,* $S_j = \emptyset$ *for any* $j \in \{1, 2, \ldots, \mu - 1\}$ *and* $|S_\mu| = 2$*;*
4. $m > 2$ *is odd,* $|S_0| = 2$*,* $|S_i| = 1$ *for some* $i \in \{1, 2, \ldots, \mu\}$ *and* $S_j = \emptyset$ *for any* $i \neq j \in \{1, 2, \ldots, \mu\}$*;*
5. $m > 2$ *is even,* $|S_0| = 2$*,* $|S_i| = 1$ *for some* $i \in \{1, 2, \ldots, \mu - 1\}$ *and* $S_j = \emptyset$ *for any* $i \neq j \in \{1, 2, \ldots, \mu\}$*;*
6. *m and n are even,* $|S_0| = 1$*,* $S_j = \emptyset$ *for any* $j \in \{1, 2, \ldots, \mu - 1\}$ *and* $|S_\mu| = 3$*;*
7. $m > 2$*, m and n are even,* $|S_0| = 1$*,* $|S_i| = 1$ *for some* $i \in \{1, 2, \ldots, \mu - 1\}$*,* $S_j = \emptyset$ *for any* $i \neq j \in \{1, 2, \ldots, \mu - 1\}$ *and* $|S_\mu| = 1$.

Proof. This lemma immediately follows from Formula (E.1). □

4 Main result

In this section we will prove necessary and sufficient conditions for a tetravalent metacirculant graph with the nonempty first symbol to be connected. Namely, we will prove the following result.

Theorem 4.1. *Let* $G = MC(m, n, \alpha, S_0, \ldots, S_\mu)$ *be a tetravalent metacirculant graph with* $S_0 \neq \emptyset$*. Then G is connected if and only if one of the following conditions holds:*

1. $m = 1$*,* $S_0 = \{\pm s, \pm r\}$ *and* $gcd(s, r, n) = 1$*;*
2. $m = 2$*, n is even,* $S_0 = \{\pm s, \frac{n}{2}\}$*,* $S_1 = \{k\}$ *and* $gcd(s, \frac{n}{2}) = 1$*;*
3. $m = 2$*,* $S_0 = \{\pm s\}$*,* $S_1 = \{k, \ell\}$ *and* $gcd(s, k - \ell, n) = 1$*;*
4. $m > 2$ *is odd,* $S_0 = \{\pm s\}$*,* $S_i = \{k\}$ *for some* $i \in \{1, 2, \ldots, \mu\}$ *such that* $gcd(i, m) = 1$*,* $S_j = \emptyset$ *for any* $i \neq j \in \{1, 2, \ldots, \mu\}$ *and* $gcd(s, r, n) = 1$*, where* $r = k(1 + \alpha^i + \cdots + \alpha^{(m-1)i})$ *reduced modulo n;*
5. $m > 2$ *is even,* $S_0 = \{\pm s\}$*,* $S_i = \{k\}$ *for some* $i \in \{1, 2, \ldots, \mu - 1\}$ *such that* $gcd(i, m) = 1$*,* $S_j = \emptyset$ *for any* $i \neq j \in \{1, 2, \ldots, \mu\}$ *and* $gcd(s, r, n) = 1$*, where* $r = k(1 + \alpha^i + \cdots + \alpha^{(m-1)i})$ *reduced modulo n;*
6. $m = 2$*, n is even,* $S_0 = \{\frac{n}{2}\}$*,* $S_1 = \{h, k, \ell\}$ *and* $gcd(h - k, k - \ell, \frac{n}{2}) = 1$*;*
7. $m > 2$ *is even, n is even* $S_0 = \{\frac{n}{2}\}$*,* $S_i = \{s\}$ *where i is odd and* $gcd(i, m) = 1$*,* $S_j = \emptyset$ *for any* $i \neq j \in \{1, 2, \ldots, \mu - 1\}$*,* $S_\mu = \{r\}$ *and* $gcd(p, \frac{n}{2}) = 1$*, where p is* $[r - s(1 + \alpha^i + \alpha^{2i} + \cdots + \alpha^{(\mu-1)i})]$ *reduced modulo n;*

8. $m > 2$ *is even, but* $\mu = \frac{m}{2}$ *is odd,* n *is even,* $S_0 = \{\frac{n}{2}\}$, $S_i = \{s\}$, *where* i *is even and* $gcd(i, m) = 2$, $S_j = \emptyset$ *for any* $i \neq j \in \{1, 2, \ldots, \mu - 1\}$, $S_\mu = \{r\}$ *and* $gcd(q, \frac{n}{2}) = 1$, *where* $i = 2^t i'$ *with* i' *to be odd and* q *to be* $\left[r(1 + \alpha^{i'} + \alpha^{2i'} + \cdots + \alpha^{(2^t-1)i'}) - s(1 + \alpha^{i'} + \alpha^{2i'} + \cdots + \alpha^{(\mu-1)i'})\right]$ *reduced modulo* n.

Proof. Let $G = MC(m, n, \alpha, S_0, \ldots, S_\mu)$ be a tetravalent metacirculant graph with $S_0 \neq \emptyset$. By Lemma 3.5, only one of Cases 1 – 7 of this lemma may occur. For each of these cases, we will consider the graph $\overline{G} = C(n, \overline{S})$ and the graph $G^0 = C(n, S^0)$ constructed from G. By Lemma 3.2, the problem of finding the necessary and sufficient conditions for G to be connected can be reduced to the same problems for both $\overline{G}$ and G^0 together. But, by Lemma 3.1, $\overline{G}$ and G^0 are circulant graphs. So we can apply Lemmas 3.3 and 3.4 to get the corresponding condition in each of these cases.

The finding $\overline{R}$ such that $\langle \overline{R} \rangle = \overline{S}$ is easy for each of these cases. In order to find R such that $\langle R \rangle = S^0$, we first specify the set R. Then we prove the equality $S^0 = \langle R \rangle$ by showing that for any walk P in G connecting a vertex v_x^0 to a vertex v_y^0 of V^0, $ch(P)$ belongs to $\langle R \rangle$. Since $ch(W_1 * W_2) = ch(W_1) + ch(W_2)$ and $ch(W * W^{-1}) = 0$, without loss of generality, we may assume that

the endvertices of P are the only vertices of V^0 in P and P has no subwalks of the type $W * W^{-1}$. $(*)$

Now we consider Cases 1 – 7 of Lemma 3.5 in turn.

(1) $|S_0| = 4$ *and* $S_1 = \cdots = S_\mu = \emptyset$.

Let $S_0 = \{\pm s, \pm r\}$. Then it is not difficult to show that $\overline{G} = C(m, \overline{S})$ with $\overline{S} = \emptyset$ and $G^0 = C(n, S^0)$ with $S^0 = \langle S_0 \rangle$. So by Lemmas 3.2 – 3.4, G is connected if and only if Case 1 of Theorem 4.1 holds.

(2) m *and* n *are even,* $|S_0| = 3$, $S_j = \emptyset$ *for any* $j \in \{1, 2, \ldots, \mu - 1\}$ *and* $|S_\mu| = 1$.

Since $0 \notin S_0 = -S_0$, S_0 must be of the form $\{\pm s, \frac{n}{2}\}$ with $s \neq 0, \frac{n}{2}$. Let $S_\mu = \{k\}$. Then $\overline{S} = \{\pm \mu\}$. We show that $S^0 = \langle S_0 \rangle$.

Let P be a walk mentioned above. If all vertices of P are in V^0 then by Assumption $(*)$, $P = v_x^0 v_{x+s}^0$ with $s \in S_0$. So $ch(P) \in \langle S_0 \rangle$. If P has vertices not in V^0, then also by Assumption $(*)$, P must be of the form $v_x^0 v_{x+k}^\mu Q v_{y+k}^\mu v_y^0$, where Q is a walk in $G[V^\mu]$. We have $ch(Q) \in \langle S_0 \rangle$ since the subgraph $G[V^\mu]$ is isomorphic to $C(n, \alpha^\mu S_0)$. So $ch(P) = k + ch(Q) - k = ch(Q) \in \langle S_0 \rangle$. Thus $S^0 = \langle S_0 \rangle$. By Lemmas 3.2 – 3.4, we conclude G is connected if and only if Case 2 of Theorem 4.1 holds.

(3) m *is even,* $|S_0| = 2$, $S_j = \emptyset$ *for any* $j \in \{1, 2, \ldots, \mu - 1\}$ *and* $|S_\mu| = 2$.

Let $S_0 = \{\pm s\}$, $S_\mu = \{k, \ell\}$ Then $\overline{G} = C(m, \overline{S})$ with $\overline{S} = \{\pm\mu\}$. By arguments similar to those of (2), we can show that $G^0 = C(n, S^0)$ with $S^0 = \langle R\rangle$, where $R = \{\pm s, \pm(k-\ell)\}$. Again by Lemmas 3.2 – 3.4, the graph G is connected if and only if Case 3 of Theorem 4.1 holds.

(4) *$m > 2$ is odd, $|S_0| = 2$, $|S_i| = 1$ for some $i \in \{1, 2, \ldots, \mu\}$ and $S_j = \emptyset$ for any $i \neq j \in \{1, 2, \ldots, \mu\}$.*

Let $S_0 = \{\pm s\}$, $S_i = \{k\}$, d be the smallest positive integer such that $di \equiv 0 \pmod{m}$ and $r = k\left(1 + \alpha^i + \cdots + \alpha^{(d-1)i}\right)$. Then it is clear that $\overline{G} = C(m, \overline{S})$ with $\overline{S} = \{\pm i\}$. We show now that $G^0 = C(n, S^0)$ with $S^0 = \langle R\rangle$, where $R = \{\pm s, \pm r\}$.

Let P be a walk mentioned in the beginning of the proof. We show that $ch(P) \in \langle R\rangle$ by induction on the number of S_0-intervals of P.

If P has no S_0-intervals, then P has to be an S_i-walk. If P is an S_i^+-walk, then P has the form $v_x^0 v_{x+k}^i v_{x+k+\alpha^i k}^{2i} \cdots v_{x+r}^0$. So $ch(P) = r \in \langle R\rangle$. If P is an S_i^--walk, then P^{-1} has the above form. Therefore, $ch(P^{-1}) \in \langle R\rangle$ and $ch(P) = -ch(P^{-1}) \in \langle R\rangle$.

If P has one S_0-interval, then either P is an S_0-walk or $P = Q_1 * Q_2 * Q_3$, where Q_1 and Q_3 are S_i-intervals and Q_2 is an S_0-interval. For the former case, it is clear that $ch(P) \in \{\pm s\} \subseteq \langle R\rangle$. For the latter case, there are the following subcases to consider:

(a) Both Q_1 and Q_3 are S_i^+-walks or both Q_1 and Q_3 are S_i^--walks.

If both Q_1 and Q_3 are S_i^+-walks, then we can write

$$Q_1 = v_x^0 v_{x+k}^i v_{x+k+\alpha^i k}^{2i} \cdots v_{x+k+\alpha^i k+\cdots+\alpha^{(t-1)i}k}^{ti} \tag{E.2}$$

$$Q_3 = v_{x'}^{ti} v_{x'+\alpha^{ti}k}^{(t+1)i} \cdots v_{x'+\alpha^{ti}k+\cdots+\alpha^{(d-1)i}k}^{di} \tag{E.3}$$

with $0 < t < d$. So

$$\begin{aligned} ch(P) &= ch(Q_1) + ch(Q_2) + ch(Q_3) \\ &= ch(Q_2) + ch(Q_1) + ch(Q_3) \\ &= ch(Q_2) + \left(k + \alpha^i k + \cdots + \alpha^{(t-1)i}k\right) + \left(\alpha^{ti}k + \cdots + \alpha^{(d-1)i}k\right) \\ &= ch(Q_2) + r \in \langle R\rangle. \end{aligned}$$

If both Q_1 and Q_3 are S_i^--walks, then $P^{-1} = Q_3^{-1} * Q_2^{-1} * Q_1^{-1}$ is the walk just considered above. So $ch(P^{-1}) \in \langle R\rangle$ and therefore $ch(P) = -ch(P^{-1})$ is also in $\langle R\rangle$.

(b) Q_1 is an S_i^+-walk and Q_3 is an S_i^--walk.

If Q_1 is as in Form (E.2), then Q_3 has to have the form

$$Q_3 = v_{x'}^{ti} v_{x'-\alpha^{(t-1)i}k}^{(t-1)i} v_{x'-\alpha^{(t-1)i}k-\alpha^{(t-2)i}k}^{(t-2)i} \cdots v_{x'-\alpha^{(t-1)i}k-\cdots-k}^0.$$

It is clear that $ch(Q_1) + ch(Q_3) = 0$. So $ch(P) = ch(Q_2) \in \langle R \rangle$.

(c) Q_1 is an S_i^--walk and Q_3 is an S_i^+-walk.

The proof is similar to that of Subcase (b).

Thus, $ch(P) \in \langle R \rangle$ if P is a walk mentioned in the beginning of the proof and has one S_0-interval.

Assume now that $t \geq 2$ and $ch(P) \in \langle R \rangle$ for all walks P' in G which have less than t S_0-intervals. Further, let P be a walk in G, which has t S_0-intervals. Then P must be of the form $P = P_1 * P_2 * \cdots * P_z$ with $P_1, P_2, \ldots, P_z$ intervals of P. By Assumption $(*)$ and $t \geq 2$, P_1 cannot be an S_0-interval. So P_2 must be an S_0-interval. Let v_b^a be the common vertex of P_2 and P_3, and W be an S_i-walk in G starting at v_b^a and terminating at a vertex of V^0. Consider the walk P'', where $P'' = P_1 * P_2 * W * W^{-1} * P_3 * \cdots * P_z$. It is easy to see that both $P_1 * P_2 * W$ and $W^{-1} * P_3 * \cdots * P_z$ are walks satisfying the induction hypothesis. So we have $ch(P_1 * P_2 * W) \in \langle R \rangle$ and $ch(W^{-1} * P_3 * \cdots * P_z) \in \langle R \rangle$. Therefore $ch(P) = ch(P'') = ch(P_1 * P_2 * W) + ch(W^{-1} * P_3 * \cdots * P_z) \in \langle R \rangle$. Thus,. $S^0 = \langle R \rangle$. By Lemmas 3.2, 3.3 and 3.4, we conclude G is connected if and only if Case 4 of Theorem 4.1 holds.

The proof of Case 5 is similar to that of Case 4 and the proof of Case 6 is similar to that of Cases 2 and 3. So we omit them here.

(7) *$m > 2$, m and n are even, $|S_0| = 1$, $|S_i| = 1$ for some $i \in \{1, 2, \ldots, \mu - 1\}$, $S_j = \emptyset$ for any $i \neq j \in \{1, 2, \ldots, \mu - 1\}$ and $|S_\mu| = 1$.*

As in Case 4, we might use here induction to prove Conditions 7 and 8 of Theorem 4.1. But there is a shorter way, which we do use here, to prove these conditions.

It is clear that $S_0 = \{\frac{n}{2}\}$. Let $S_i = \{s\}$ and $S_\mu = \{r\}$. Denote $w_j^i = \{v_j^i, v_{j+\frac{n}{2}}^i\}$, where $i \in \mathbb{Z}_m$ and $j \in \mathbb{Z}_{n/2}$. We construct the graph G' from G as follows. The vertex-set $V(G')$ is the set $\{w_j^i \mid i \in \mathbb{Z}_m, j \in \mathbb{Z}_{n/2}\}$. Two vertices w_j^i and w_ℓ^k are adjacent in G' if and only if there are $u \in w_j^i$ and $v \in w_\ell^k$ such that they are adjacent in G. It is not difficult to verify that G' is isomorphic to the metacirculant graph $MC(m, \frac{n}{2}, \alpha', S_0', \ldots, S_\mu')$, where $\alpha' \equiv \alpha \pmod{\frac{n}{2}}$, $S_i' = \{s'\}$ with $s' \equiv s \pmod{\frac{n}{2}}$, $S_j' = \emptyset$ for $i \neq j \in \{0, 1, \ldots, \mu - 1\}$ and $S_\mu' = \{r'\}$ with $r' \equiv r \pmod{\frac{n}{2}}$. So we can identify G' with this metacirculant graph. Therefore, G' is a cubic metacirculant graph with $S_0' = \emptyset$. Also, it is not difficult to show that in this case G is connected if and only if G' is connected. By Theorem 2 in [4], G' is connected if and only if one of the following conditions holds:

(i) $S_i' = \{s'\}$ where i is odd and $\gcd(i, m) = 1$, $S_j' = \emptyset$ for any $i \neq j \in \{1, 2, \ldots, \mu - 1\}$, $S_\mu' = \{r'\}$ and $\gcd(p', \frac{n}{2}) = 1$, where p' is $[r' - s'(1 + (\alpha')^i +$

$(\alpha')^{2i} + \cdots + (\alpha')^{(\mu-1)i})]$ reduced modulo $\frac{n}{2}$;

(ii) $\mu = \frac{m}{2}$ is odd, $S'_i = \{s'\}$ where i is even and $\gcd(i, m) = 2$, $S'_j = \emptyset$ for any $i \neq j \in \{1, 2, \ldots, \mu - 1\}$, $S'_\mu = \{r'\}$ and $\gcd(q', \frac{n}{2}) = 1$, where $i = 2^t i'$ with i' odd and q' to be $\left[r'(1 + (\alpha')^{i'} + (\alpha')^{2i'} + \cdots + (\alpha')^{(2^t-1)i'}) - s'(1 + (\alpha')^{i'} + (\alpha')^{2i'} + \cdots + (\alpha')^{(\mu-1)i'})\right]$ reduced modulo $\frac{n}{2}$.

From Conditions (i) and (ii) above for G', we can easily get Conditions 7 and 8 of Theorem 4.1, respectively, for G, because $s' \equiv s \pmod{\frac{n}{2}}$, $\alpha' \equiv \alpha \pmod{\frac{n}{2}}$ and $r' \equiv r \pmod{\frac{n}{2}}$.

The proof of Theorem 4.1 is complete. □

Theorem 4.1 gives us an algorithm to test connectedness of a tetravalent metacirculant graph $G = MC(m, n, \alpha, S_0, \ldots, S_\mu)$ with $S_0 \neq \emptyset$ by verifying if the parameters m, n, α, S_0, ... , S_μ of G satisfy one of Conditions 1 – 8 of Theorem 4.1.

Acknowledgement

We would like to express our sincere thanks to the referees for many valuable comments and useful suggestions which help us to improve the paper.

References

[1] B. Alspach and T. D. Parsons, "A construction for vertex transitive graphs", *Canad. J. Math.* 34 (1982), 307 – 318.

[2] M. Behzad and G. Chartrand, *Introduction to the theory of graphs*, Allyn and Bacon, Boston (1971).

[3] Ngo Dac Tan, "Hamilton cycles in cubic $(4, n)$-metacirculant graphs", *Acta Math. Vietnamica* 17 (1992), 83 – 93.

[4] Ngo Dac Tan, "Connectedness of cubic metacirculant graphs", *Acta Math. Vietnamica* 18 (1993), 3 – 17.

[5] Ngo Dac Tan, "Hamilton cycles in cubic (m, n)-metacirculant graphs with m divisible by 4", *Graphs and Combin.* 10 (1994), 67 – 73.

[6] Ngo Dac Tan, "Non-Cayley tetravalent metacirculant graphs and their hamiltonicity", *Journal of Graph Theory* 23 (1996), 273 – 287.

[7] Ngo Dac Tan and Tran Minh Tuoc, "On Hamilton cycles in connected tetravalent metacirculant graphs with nonempty first symbol", *Acta Math. Vietnamica* (accepted).

[8] Ngo Dac Tan and Tran Minh Tuoc, "Connectedness of tetravalent metacirculant graphs with the empty first symbol", *Preprint 2002/33*, Hanoi Institute of Mathematics (2002) (submitted).

[9] H. Wielandt, *Finite permutation groups*, Academic Press, New York (1964).

On the Relation Between Maximum Entropy Principle and the Condition Independence Assumption in the Probabilistic Logic

Ha Dang Cao Tung

Information Technology Department, Hanoi Teacher Training College

Abstract. In the Probabilistic Logic, the single value of the truth probability of a sentence is calculated by adding assumptions. By adding the Maximum Entropy Principle (MEP) the reasoning problem becomes a nonlinear optimization problem. In this note, we assert the equivalence between the MEP and the Conditional Independence assumption (CIA) and the Randomness Assumtpion (RA) formulas for a class of Probabilistic Knowledge Base. Thus, for this class, instead of solving a nonlinear optimization problem to find the truth probability of a sentence, we can drive it from Calculus $\mathcal{C}$ and some CIA + RA formulas.

1. Introduction

Given a set of sentences $\Gamma = \{S_1, S_2, ..., S_L\}$ and a sentence S. Denote by $\Delta(\Gamma, S)$ (or Δ shortly) the set of propositional variables $\{A_1, A_2, ..., A_n\}$ occurring in $\{S_1, S_2, ..., S_L, S\}$. Each assignment of boolean values to the variables in Δ defines a possible world [4]. For every given probability distribution $\mathbf{p} = (p_1, p_2, ..., p_N)$, where $N = 2^n$, on the set of possible worlds, the truth probability of a sentence S is defined to be the sum of probabilities of possible worlds in which S is true.

In Probabilistic Logic (PL), from any knowledge base (KB) given by the truth probabilities $p(S_1), p(S_2), ..., p(S_L)$ of setences $S_1, S_2, ..., S_L$ one can usually determinate the lower and upper bounds of an interval containing the truth probability $p(S)$, but not the single value of $p(S)$. A such single value of $p(S)$ can be determined from probabilities $p(S_1), p(S_2), ..., p(S_L)$ only for special cases. In [1], it was proved that the truth probability of S was calculated from the truth probabilities of $S_1, S_2, ..., S_L$ by a formula

$$p(S) = f(p(S_1), ..., p(S_L)) \tag{2}$$

if and only if (1) is deducible in the Calculus $\mathcal{C}$ comprising of two axioms:

1. $p(P) = 1$ if P is tautology.
2. $p(P \vee Q) = p(P) + p(Q)$ if $\neg(P \wedge Q)$ is tautology.

In other words, if $p(S)$ can be determined uniquely from $p(S_1), p(S_2), ...,$ $p(S_L)$, then it is deducible in Calculus $\mathcal{C}$. In the general case $p(S)$ can not be determined uniquely from $p(S_1), p(S_2), ..., p(S_L)$, for finding a unique value for $p(S)$ one has to add some assumption into KB, e.g., the maximum entropy principle (MEP) [4]. In PL with MEP [4] MEP is used to calculate a probalbility distribution $\mathbf{p} = (p_1, p_2, ..., p_N)$ on the set of possible worlds, and then, from this distribution $\mathbf{p}$ one can determine a single value of $p(S)$, this value is a function of $p(S_1)$, $p(S_2), ..., p(S_L)$, and is denoted by:

$$p(S) = f_{\text{MEP}}(p(S_1), p(S_2), ..., p(S_L)). \tag{2}$$

An another assumption, the conditinal independence assumption (CIA) is considered in [1]. The CIA is stated in the form "two sentences A and B are independent with respects to C" and is expressed by the formula

$$p(A \wedge B|C) = p(A|C) \times p(B|C)$$

(when C is a tautology we have a particular case $p(A \wedge B) = p(A) \times p(B)$, A and B are independent). Together with the CIA one considers also a supplementary assumption called the randomness assumption (RA) which is usually expressed in the form

$$p(A|B) = \frac{1}{2}.$$

(under the condition B, A is random).

In probability theory, the MEP is often considered together with CIA. In PL, the similar problem on the relation betwen MEP and CIA is posed in [1] as follows: Given a knowledge base with the set of sentences $\Gamma = \{S_1, S_2, ..., S_L\}$ and a sentence S. It is possible to determine a set H of CIA and RA formulas such that (2) is derived in Calculus $\mathcal{C} + H$? In this note we shall the possitive answer to this problem for a class of probabilistic knowledge bases.

Remark. In [4], the possible workds are defined by the consistent vectors of values of $\{S_1, S_2, ..., S_L, S\}$, and on the basis of this definition one can obtain "inconsistent" results as shown in the following example: Let $\mathbf{B}$ be the KB given $\mathbf{B} = \{\langle AB, \alpha \rangle\}$, where $0 \leq \alpha < 1$; from this KB by the method used in [4] can we can calculate separately the truth probabilities of $A\overline{B}, \overline{A}B, \overline{A}\,\overline{B}$:

$$p(A\overline{B}) = p(\overline{A}B) = p(\overline{A}\,\overline{B}) = \frac{1}{2}(1-\alpha),$$

hence we obtain

$$p(AB) + p(A\overline{B}) + p(\overline{A}B) + p(\overline{A}\,\overline{B}) = 1 + \frac{1}{2}(1-\alpha) > 1,$$

that violates an axiom of the probability theory. Therefore, in this paper we consider a possible workd of a set Γ of sentence as an assignment of boolean values to the propositonal variables occuring in Γ.

2. The main result

Let us give a set of sentances $\Gamma = \{S_1, S_2, ..., S_L\}$. We say that Γ is *complete* if each assigment of the boolean values to the sentences of Γ is consistent. We denote by Γ^u the formula $S_1^{u_1} S_2^{u_2} ... S_L^{u_L} (S_i^1 = S_i, S_i^0 = \overline{S_i})$, where $u = (u_1, u_2, ..., u_L)$ is a boolean vector.

Suppose that Γ can be partitioned into m subsets $\Gamma_1, ..., \Gamma_m$, where

$$\Gamma_k = \left\{S_1^{(k)}, S_2^{(k)}, ..., S_{n_k}^{(k)}\right\}, \quad k = \overline{1, m},$$

such that $(\forall k \neq l)$ $\left(\forall i = \overline{1,\ n_k};\ \ \forall j = \overline{1 n_k}\right) \neg (S_i^{(k)} S_j^{(l)})$ = Tautology. Then we say that Γ *is partitionable into* m *mutual excusive subsets.*

Theorem. *Suppose that* Γ *is partitionable into mutual exclusive subsets* $\Gamma_1, ..., \Gamma_m$*, and each of these subsets is complete. Let* H *is the set of CIA formulas of the form:*

$$p\Big(\bigwedge_{i \in I} S_i^{(k)} \Big| \bigwedge_{\substack{l=1 \\ l \neq k}}^{m} \big(\bigwedge_{j \in \overline{n_l}} \overline{S_j^{(l)}} \big)\Big) =$$

$$\prod_{i \in I} p\Big(S_i^{(k)} \Big| \bigwedge_{\substack{l=1 \\ l \neq k}}^{m} \big(\bigwedge_{j \in \overline{n_l}} \overline{S_j^{(l)}} \big)\Big) \quad (\forall k = \overline{1, m}) \ (\forall I \subseteq \overline{n_k}), \tag{3}$$

where, $\overline{n_l} = \{1, 2, ..., n_l\}$*. Then, for any KB with the set of sentences* Γ *and any sentence* S*, the values of* $p(S)$ *deduced from* $C + H$ *and from* $PL + MEP$ *are the same.*

In order to prove the theorem, we first prove the following

Lemma. *Suppose that*

$$\sum_{k=1}^{m} \Big(\max_{1 \leq i \leq n_k} (\alpha_i^{(k)}) \Big) \leq 1, \tag{4}$$

where $0 \leq \alpha_i^{(k)} \leq 1 (\forall k = 1, ..., m)$ $(\forall i = 1, ..., n_k)$*. Then the following system of equations for* $t, \xi_1, \xi_2, ..., \xi_m$

$$\begin{cases} \sum\limits_{k=1}^{m} \xi_k & = 1 \\ \xi_k \prod\limits_{i=1}^{n_k} \Big(1 - \dfrac{\alpha_i^{(k)}}{\xi_k}\Big) & = t, \quad k = 1, ..., m \end{cases} \tag{5}$$

has a unique solution.

Proof. For each $k = 1, ..., m$, we consider the function:

$$f_k(x) = x \times \prod_{i=1}^{n_k} \left(1 - \frac{\alpha_i^{(k)}}{x}\right)$$

It is easy to see that $f_k(x)$ is continuous and increasing. Therefore the inverse function $f_k^{-1}(t)$ exists, is continuous and also increasing.

For $t_1 = 0$,

$$\forall k, \quad f_k\left(\max_{i \in \overline{n_k}} \left(\alpha_i^{(k)}\right)\right) = 0 \Rightarrow f_k^{-1}(0) = \max_{i \in \overline{n_k}} \left(\alpha_i^{(k)}\right). \tag{6}$$

For $t_2 = \max_{k=\overline{1,m}} \left(\prod_{i \in \overline{n_k}} (1 - \alpha_i^{(k)}) \right)$,

$$\prod_{i \in \overline{n_k}} (1 - \alpha_i^{(k)}) \leq t_2 \Rightarrow 1 = f_k^{-1}\left(\prod_{i \in \overline{n_k}} (1 - \alpha_i^{(k)})\right) \leq f_k^{-1}(t_2). \tag{7}$$

For the function

$$F(t) = \sum_{k=1}^{m} f_k^{-1}(t),$$

we have the following equalities:

$$F(t_1) = F(0) \overset{(6)}{=} \sum_{k=1}^{m} \left(\max_{i \in \overline{n_k}} \left(\alpha_i^{(k)}\right)\right) \overset{(4)}{\leq} 1 \leq m \overset{(7)}{\leq} \sum_{k=1}^{m} f_k^{-1}(t_2) = F(t_2)$$

From the continuity of F, we have $\exists t^* | F(t^*) = 1$. Put $f_k^{-1}(t^*) = \xi_k$, it is easy to see that the system (5) has solution $(\xi_1, \xi_2, ..., \xi_m, t^*)$. We notice that the left side of the first equation in (5) is an increasing function of t. Therefore the found solution is unique.

Proof of the theorem. Suppose that Γ is partitioned into complete subsets $\Gamma_1 \ldots, \Gamma_m$, where

$$\Gamma_k = \left\{S_1^{(k)}, S_2^{(k)}, ..., S_{n_k}^{(k)}\right\}, \quad k = \overline{1,m},$$

such that $(\forall k \neq l)$ $(\forall i = \overline{1, n_k};\ \forall j = \overline{1, n_l}) \neg(S_i^{(k)} S_j^{(l)}) =$ Tautology. Let **B** be a KB with the set of sentences Γ given by $p(S_i) = \alpha_i$ for every $S_i \in \Gamma$. Then we have (see [5]):

$$\begin{cases} \sum\limits_{k=1}^{m} \left(\sum\limits_{i=1}^{2^{n_k}} p_i^{(k)} \right) & = 1 \\ \sum\limits_{i=1}^{2^{n_k}} u_{ji}^{(k)} p_i^{(k)} & = \alpha_j^{(k)}, \quad (k = \overline{1,m})(j = \overline{1,n_k}) \end{cases}$$

Put in these equations

$$p_i^{(k)} a_0 \prod_{u_{ji}^{(k)}=1} a_j^{(k)}, \quad (k = \overline{1,m})\ (i = 1, ..., 2^{n_k}),$$

we obtain

$$\begin{cases} a_0\Big[\sum_{I\subseteq\overline{n_1}}\Big(\prod_{i\in I} a_i^{(1)}\Big) + \sum_{I\subseteq\overline{n_2}}\Big(\prod_{i\in I} a_i^{(2)}\Big) + ... + \sum_{I\subseteq\overline{n_m}}\Big(\prod_{i\in I} a_i^{(m)}\Big)\Big] = 1 \\ a_0 a_j^{(k)} \sum_{I\subseteq\overline{n_k}\setminus\{j\}}\Big(\prod_{i\in I} a_i^{(k)}\Big) = \alpha_j^{(k)}, \quad (k = \overline{1,m})\ (j = \overline{1,n_k}) \end{cases}$$

where $\overline{n_k} = \{1, 2, ..., n_k\}$.

By the lemma, there are the unique values of a_0 and ξ_k such that

$$\begin{cases} \sum_{k=1}^{m} \xi_k & = 1 \\ \xi_k \prod_{i=1}^{n_k}\Big(1 - \frac{\alpha_i^{(k)}}{\xi_k}\Big) & = a_0, \quad k = 1, ..., m \end{cases}$$

For each $k = 1, ..., m$, we consider system of equations

$$\begin{cases} a_0 \sum_{I\subseteq\overline{n_k}}\big(\prod_{i\in I} a_i^{(k)}\big) & = \xi_k \\ a_0 a_j^{(k)} \sum_{I\subseteq\overline{n_k}\setminus\{j\}}\big(\prod_{i\in I} a_i^{(k)}\big) & = a_j^{(k)}, \quad (j = \overline{1,n_k}) \end{cases}$$

Put

$$\frac{a_0}{\xi_k} = a_0^{(k)}.$$

By dividing both sides of each equation of this system by ξ_k, we obtain the following system $(k = 1, ..., m)$:

$$\begin{cases} a_0^{(k)}\big(a_j^{(k)} + 1\big) \sum_{I\subseteq\overline{n_k}\setminus\{j\}}\Big(\prod_{i\in I} a_i^{(k)}\Big) & = 1 \\ a_0^{(k)} a_j^{(k)} \sum_{I\subseteq\overline{n_k}\setminus\{j\}}\Big(\prod_{i\in I} a_i^{(k)}\Big) & = \beta_j^{(k)}\Big(= \frac{\alpha_j^{(k)}}{\xi_k}\Big), \quad (j = \overline{1,n_k}) \end{cases}$$

This system of equations has the solution

$$\begin{cases} a_0^{(k)} = & \prod_{j\in\overline{n_k}}\big(1 - \beta_j^{(k)}\big) \\ a_j^{(k)} = & \frac{\beta_j^{(k)}}{a - \beta_j^{(k)}}, \quad (j = \overline{1,n_k}) \end{cases}$$

Hence, we obtain

$$p_i^{(k)} = a_0 \prod_{u_{ji}^{(k)}=1} a_j^{(k)} = \xi_k \times \prod_{u_{ji}^{(k)}=1} \beta_j^{(k)} \times \prod_{u_{ji}^{(k)}=0} (1-\beta_j^{(k)}), \quad (k=\overline{1,m}) \ (i=1,...,2^{n_k}) \tag{8}$$

Now we will show that the values (8) is exactly the values for $p_i^{(k)}$ obtained from $\mathcal{C}$ and the formulas (3). Put

$$\begin{cases} Q_k = & \bigwedge_{\substack{l=1 \\ l\neq k}}^{m} \Big(\overline{\bigwedge_{j\in\overline{n_l}} S_j^{(l)}} \Big) = \bigwedge_{\substack{j\in\overline{n_l} \\ l=\overline{1,m} \\ l\neq k}} \overline{S_j^{(l)}} \\ p(Q_k) = & \xi_k, \end{cases} \tag{9}$$

we have

$$\forall k=\overline{1,m}, \forall i\in\overline{n_k}, \ S_i^{(k)} = S_i^{(k)} Q_k \vee S_i^{(k)} \overline{Q_k} = S_i^{(k)} Q_k \vee S_i^{(k)} \overline{\overline{\bigwedge_{\substack{j\in n_l \\ l=\overline{1,m} \\ l\neq k}} \overline{S_j^{(l)}}}}$$

$$= S_i^{(k)} Q_k \vee S_i^{(k)} \bigvee_{\substack{j\in\overline{n_l} \\ l=\overline{1,m} \\ l\neq k}} S_j^{(l)}$$

$$= S_i^{(k)} Q_k \vee \bigvee_{\substack{j\in\overline{n_l} \\ l=\overline{1,m} \\ l\neq k}} (S_i^{(k)} S_j^{(l)})$$

$$= S_i^{(k)} Q_k.$$

$$p(S_i^{(k)}|Q_k) = \frac{p(S_i^{(k)} Q_k)}{p(Q_k)} = \frac{(S_i^{(k)})}{p(Q_k)} = \frac{p(S_i^{(k)})}{\xi_k} = \frac{\alpha_i^{(k)}}{\xi_k} = \beta_i^{(k)}$$

From (3) and (9) we have

$$\forall k=\overline{1,m}, \quad \forall I\subseteq\overline{n_k}, \quad p(\bigwedge_{i\in I} S_i^{(k)}|Q_k) = \prod_{i\in I} p(S_i^{(k)}|Q_k).$$

Then for all consistent vector $u_i^{(k)}$, we obtain

$$p_i^{(k)} = p(\Gamma^{u_i^{(k)}}) = p\Big(Q_k \bigwedge_{j\in\overline{n_k}} (S_j^{(k)})^{u_{ji}^{(k)}}\Big)$$

$$= p(Q_k) p\Big(\bigwedge_{j\in\overline{n_k}} (S_j^{(k)})^{u_{ji}^{(k)}} |Q_k\Big) = p(Q_k) \prod_{j\in\overline{n_k}} p\Big[(S_j^{(k)})^{u_{ji}^{(k)}} |Q_k\Big]$$

$$= \xi_k \prod_{\substack{j\in\overline{n_k} \\ u_{ji}^{(k)}=1}} p(S_j^{(k)}|Q_k) \times \prod_{\substack{j\in\overline{n_k} \\ u_{ji}^{(k)}=0}} p(\overline{S_j^{(k)}}|Q_k) = \xi_k \prod_{\substack{j\in\overline{n_k} \\ u_{ji}^{(k)}=1}} \beta_j^{(k)} \times \prod_{\substack{j\in\overline{n_k} \\ u_{ji}^{(k)}=0}} (1-\beta_j^{(k)})$$

Thus, we obtain again the formula (8). The theorem is proved.

Remark. *If* $\Gamma = \{S_1, S_2, ..., S_L\}$ *is itself a complete set of sentences, then the set* H *stated in the theorem consists of the following CIA formulas:*

$$p(\bigwedge_{i\in I} S_i) = \prod_{i\in I} p(S_i), \quad \forall I \subseteq \{1, 2, ..., L\},$$

We notice in the reasoning, if a propositional variable A occurs in the goal S and does not occur in Γ, we have to add to H the RA formula:

$$p(A) = \frac{1}{2}.$$

In general, if v and w are two possible worlds having the same consistent vector for Γ, we have to add to H the RA formula:

$$p(\Delta^v | \Delta^v \vee \Delta^w) = \frac{1}{2}. \tag{10}$$

Example. Given a KB consisting of two formulas $p(C) = \alpha$, $p(AC \vee B) = \beta$. Let $S_1 = C$, $S_2 = AC \vee B$. The set $\{C, AC \vee B\}$ is complete. Then H contains the following formulas

$$p(S_1.S_2) = p(S_1)p(S_2),$$
$$p(B|AC) = p(A|BC) = p(A|\overline{B}\,\overline{C}) = p(A|B\overline{C}) = \frac{1}{2}.$$

For $S = AB$, from $C + H$ we have

$$\left.\begin{array}{l} p(B|AC) = \frac{1}{2} \Rightarrow p(ABC) = \frac{1}{2}p(AC) \Rightarrow p(ABC) = p(A\overline{B}C) \\ p(A|BC) = \frac{1}{2} \Rightarrow p(ABC) = \frac{1}{2}p(BC) \Rightarrow p(ABC) = p(\overline{A}BC) \end{array}\right\} \Rightarrow$$

$$\Rightarrow p(ABC) = \frac{1}{3}p((ABC) + p(A\overline{B}C) + p(\overline{A}BC)) = \frac{1}{3}p(ABC \vee A\overline{B}C \vee \overline{A}BC)$$
$$= \frac{1}{3}\, p((A \vee B)C)$$
$$p(A|B\overline{C}) = \frac{1}{2} \Rightarrow p(AB\overline{C}) = \frac{1}{2}p(B\overline{C})$$

Hence,

$$\begin{aligned}p(AB) &= p(ABC \vee AB\overline{C}) = p(ABC) + p(AB\overline{C})\\ &= \frac{1}{3}p((A \vee B)C) + \frac{1}{2}p(B\overline{C}) = \frac{1}{3}p(C(AC \vee B)) + \frac{1}{2}p(\overline{C}(AC \vee B))\\ &= \frac{1}{3}p(S_1S_2) + \frac{1}{2}p(\overline{S_1}S_2) = \frac{1}{3}p(S_1)p(S_2) + \frac{1}{2}p(\overline{S_1})p(S_2)\\ &= \frac{1}{6}\beta(3-\alpha)\end{aligned}$$

This result can be obtained also by a calculation from PL + MEP.

3. Conclusion

Thus, in this paper we have obtained the positive answer to the question of equivalence between the MEP and the CIA + RA for a class of probabilistic knowledge bases. And therefore, for this class we can deduce the truth probability of any sentence from the Calcules C and a suitable set H of CIA and RA formulas instead of solving a nonlinear optimization problem.

References

1. Dieu P. D., On a Theory of Interval-Valued Probabilistic Logic. *Research Report* (Vietnam NCST, 1991).
2. Guggeneimer H. and Feedman R. S., Foundation of probabilistic Logic. *Proceedings of 10th International Joint Conference on Artificial Intelligence* (1987) 939-941.
3. McLeish M., A note on probabilistic logic. *Proceedings AIIC* (1988) 215-219.
4. Nilsson N. J., Probabilistic logic. *Artificial Intelligence* **28** (1986) 71-87.
5. Tung H. D. C., Interval-Valued probabilistic Logic for a Class of Horn Clauses. *Vietnam J. Math.* **25**: 3 (1997) 241-252.